NATIONAL ACADEMIES
Sciences
Engineering
Medicine

NATIONAL ACADEMIES PRESS
Washington, DC

Carbon Dioxide Utilization Markets and Infrastructure

Status and Opportunities

A First Report

Committee on Carbon Utilization Infrastructure, Markets, Research and Development

Board on Energy and Environmental Systems

Division on Engineering and Physical Sciences

Board on Chemical Sciences and Technology

Division on Earth and Life Studies

Consensus Study Report

NATIONAL ACADEMIES PRESS 500 Fifth Street, NW Washington, DC 20001

This activity was supported by Contract DE-EP0000026/89303021FFE400026 with the U.S. Department of Energy. Any opinions, findings, conclusions, or recommendations expressed in this publication do not necessarily reflect the views of any organization or agency that provided support for the project.

International Standard Book Number-13: 978-0-309-69327-1
International Standard Book Number-10: 0-309-69327-6
Digital Object Identifier: https://doi.org/10.17226/26703

This publication is available from the National Academies Press, 500 Fifth Street, NW, Keck 360, Washington, DC 20001; (800) 624-6242 or (202) 334-3313; http://www.nap.edu.

Copyright 2023 by the National Academy of Sciences. National Academies of Sciences, Engineering, and Medicine and National Academies Press and the graphical logos for each are all trademarks of the National Academy of Sciences. All rights reserved.

Printed in the United States of America.

Suggested citation: National Academies of Sciences, Engineering, and Medicine. 2023. *Carbon Dioxide Utilization Markets and Infrastructure: Status and Opportunities: A First Report*. Washington, DC: The National Academies Press. https://doi.org/10.17226/26703.

The **National Academy of Sciences** was established in 1863 by an Act of Congress, signed by President Lincoln, as a private, nongovernmental institution to advise the nation on issues related to science and technology. Members are elected by their peers for outstanding contributions to research. Dr. Marcia McNutt is president.

The **National Academy of Engineering** was established in 1964 under the charter of the National Academy of Sciences to bring the practices of engineering to advising the nation. Members are elected by their peers for extraordinary contributions to engineering. Dr. John L. Anderson is president.

The **National Academy of Medicine** (formerly the Institute of Medicine) was established in 1970 under the charter of the National Academy of Sciences to advise the nation on medical and health issues. Members are elected by their peers for distinguished contributions to medicine and health. Dr. Victor J. Dzau is president.

The three Academies work together as the **National Academies of Sciences, Engineering, and Medicine** to provide independent, objective analysis and advice to the nation and conduct other activities to solve complex problems and inform public policy decisions. The National Academies also encourage education and research, recognize outstanding contributions to knowledge, and increase public understanding in matters of science, engineering, and medicine.

Learn more about the National Academies of Sciences, Engineering, and Medicine at **www.nationalacademies.org**.

Consensus Study Reports published by the National Academies of Sciences, Engineering, and Medicine document the evidence-based consensus on the study's statement of task by an authoring committee of experts. Reports typically include findings, conclusions, and recommendations based on information gathered by the committee and the committee's deliberations. Each report has been subjected to a rigorous and independent peer-review process and it represents the position of the National Academies on the statement of task.

Proceedings published by the National Academies of Sciences, Engineering, and Medicine chronicle the presentations and discussions at a workshop, symposium, or other event convened by the National Academies. The statements and opinions contained in proceedings are those of the participants and are not endorsed by other participants, the planning committee, or the National Academies.

Rapid Expert Consultations published by the National Academies of Sciences, Engineering, and Medicine are authored by subject-matter experts on narrowly focused topics that can be supported by a body of evidence. The discussions contained in rapid expert consultations are considered those of the authors and do not contain policy recommendations. Rapid expert consultations are reviewed by the institution before release.

For information about other products and activities of the National Academies, please visit www.nationalacademies.org/about/whatwedo.

COMMITTEE ON CARBON UTILIZATION INFRASTRUCTURE, MARKETS, RESEARCH AND DEVELOPMENT

EMILY A. CARTER (NAS/NAE), Princeton University and Princeton Plasma Physics Laboratory, *Chair*
SHOTA ATSUMI, University of California, Davis
MAKINI BYRON, Linde
ALAYNA CHUNEY,[1] Carbon180
STEPHEN COMELLO, Stanford Graduate School of Business and EFI Foundation
MAOHONG FAN, University of Wyoming and Georgia Institute of Technology
MATTHEW FRY, Great Plains Institute
HAROUN MAHGEREFTEH, University College London
EMANUELE MASSETTI, Georgia Institute of Technology
AH-HYUNG (ALISSA) PARK, Columbia University
JOSEPH B. POWELL (NAE), Shell (retired)
ANDREA RAMÍREZ RAMÍREZ, Delft University of Technology
VOLKER SICK, University of Michigan

Staff

ELIZABETH ZEITLER, Associate Director, Board on Energy and Environmental Systems (BEES), *Study Co-Director*
CATHERINE WISE, Program Officer, BEES, *Study Co-Director*
LIANA VACCARI, Program Officer, Board on Chemical Sciences and Technology
REBECCA DeBOER, Research Associate, BEES
JASMINE BRYANT, Research Assistant, BEES
KAIA RUSSELL, Program Assistant, BEES

NOTE: See Appendix B, Disclosure of Conflicts of Interest.
[1] Resigned May 2022.

BOARD ON ENERGY AND ENVIRONMENTAL SYSTEMS

JARED COHON (NAE), Carnegie Mellon University, *Chair*
VICKY BAILEY, Anderson Stratton Enterprises, LLC; BHMM Energy Services, LLC
CARLA BAILO, Center for Automotive Research
DEEPAKRAJ DIVAN (NAE), Georgia Institute of Technology
MARCIUS EXTAVOUR, XPRIZE Foundation
T.J. GLAUTHIER, TJG Energy Associates, LLC
PAULA GLOVER, Alliance to Save Energy
AMOS GOLDHABER, Claremont Creek Ventures
DENISE GRAY (NAE), LG Chem Michigan Inc. Tech Center
JENNIFER HOLMGREN, LanzaTech
JOHN KASSAKIAN (NAE), Massachusetts Institute of Technology
MICHAEL LAMACH, Trane Technologies (retired)
JOSÉ SANTIESTEBAN (NAE), ExxonMobil Research and Engineering Company (retired)
ALEXANDER SLOCUM, SR. (NAE), Massachusetts Institute of Technology
SUSAN TIERNEY, Analysis Group
GORDON VAN WELIE (NAE), ISO New England, Inc.
DAVID VICTOR, University of California, San Diego, Deep Decarbonization Initiative

Staff

K. JOHN HOLMES, Director/Scholar
ELIZABETH ZEITLER, Associate Director
BRENT HEARD, Program Officer
KASIA KORNECKI, Program Officer
CATHERINE WISE, Program Officer
REBECCA DeBOER, Research Associate
KYRA HOWE, Research Assistant
JASMINE BRYANT, Research Assistant
KAIA RUSSELL, Program Assistant
HEATHER LOZOWSKI, Financial Manager

BOARD ON CHEMICAL SCIENCES AND TECHNOLOGY

SCOTT COLLICK, DuPont, *Co-Chair*
JENNIFER SINCLAIR CURTIS, University of California, Davis, *Co-Chair*
GERARD BAILLELY, Procter & Gamble Company
RUBEN CARBONELL (NAE), North Carolina State University
JOHN FORTNER, Yale University
KAREN GOLDBERG (NAS), Vagelos Institute for Energy Science and Technology, University of Pennsylvania
JENNIFER HEEMSTRA, Emory University
JODIE LUTKENHAUS, Texas A&M University
SHELLEY MINTEER, University of Utah
AMY PRIETO, Colorado State University and Prieto Battery, Inc.
MEGAN ROBERTSON, University of Houston
SALY ROMERO-TORRES, Thermo Fisher Scientific Pharma Services
REBECCA RUCK, Merck Research Laboratories
ANUP K. SINGH, Lawrence Livermore National Laboratory
VIJAY SWARUP, ExxonMobil Research and Engineering Company

Staff

CHARLES FERGUSON, Director
LIANA VACCARI, Program Officer
LINDA NHON, Program Officer
JESSICA WOLFMAN, Research Associate
BRENNA ALBIN, Senior Program Assistant
AYANNA LYNCH, Research Assistant
KAYANNA WYMBS, Program Assistant
NICHOLAS ROGERS, Senior Finance Business Partner
THANH NGUYEN, Finance Business Partner

Preface

As we move further into the third decade of the twenty-first century, the world continues to witness ever-more concerning indicators of global climate change, from year-round wildfires of unprecedented size to megadroughts to massive flooding, exacerbated by the burning of fossil carbon that has powered our civilization for centuries. The challenge is clear and urgent: How do we maintain or improve quality of life for the planet's inhabitants while ameliorating the harm already done and preventing future harm to the environment? One essential part of the strategy has to be to stop, on a global net basis, emitting gases to the atmosphere that warm Earth, especially but not exclusively carbon dioxide, because of its relatively high concentration and long life in the atmosphere.

A global transition to net-zero greenhouse gas emissions, necessary for maintaining a safe, stable climate, will require overcoming technological and societal challenges. A key component in achieving net-zero emissions is carbon management, which involves mitigating the vast majority of carbon dioxide emissions and ensuring that remaining flows of carbon dioxide to and from the atmosphere are balanced. Carbon dioxide utilization, the focus of this report, can play a productive role in achieving net-zero emissions by providing pathways for carbon storage or carbon removal in useful products in some cases and by enabling a circular carbon economy in others. Long-lived products, such as concrete and aggregates, can store carbon originating from fossil-derived emissions or, if produced from atmospheric or other sustainable sources of carbon dioxide, can durably remove carbon dioxide from the environment. A circular carbon economy will allow continued production and use of carbon-based products, such as aviation fuels, building materials, plastics, and commodity chemicals, without releasing net carbon dioxide emissions to the atmosphere. This first report from the Committee on Carbon Utilization Infrastructure, Markets, Research and Development identifies priority options for carbon dioxide–derived products that could participate in a future net-zero-emission economy, discusses the associated infrastructure requirements and deployment opportunities, and explores policy, regulatory, and societal considerations.

To address this wide breadth of topics, the National Academies of Sciences, Engineering, and Medicine convened a committee with diverse expertise and experience, ranging from technology research and development to industrial gas and chemicals processing, to pipeline development and operations, to policy, societal, environmental, and economic analysis. The committee has worked tirelessly over the past 7 months, holding public webinars to gather information from experts and engaging in rigorous yet respectful discussions, to produce a report that reflects these various perspectives and provides valuable insights into opportunities for carbon dioxide utilization. I would like to thank all of the committee members for their commitment to this project and the National Academies staff for their outstanding support, and I look forward to working with them through the remainder of the study.

Emily A. Carter, *Chair*
Committee on Carbon Utilization Infrastructure, Markets,
Research and Development

Reviewers

This Consensus Study Report was reviewed in draft form by individuals chosen for their diverse perspectives and technical expertise. The purpose of this independent review is to provide candid and critical comments that will assist the National Academies of Sciences, Engineering, and Medicine in making each published report as sound as possible and to ensure that it meets the institutional standards for quality, objectivity, evidence, and responsiveness to the study charge. The review comments and draft manuscript remain confidential to protect the integrity of the deliberative process.

We thank the following individuals for their review of this report:

JOHN BENEMANN, MicroBio Engineering
ANDRÉ L. BOEHMAN, University of Michigan
MARCIUS EXTAVOUR, XPRIZE Foundation
CHRIS GREIG, Princeton University
COMAS HAYNES, Georgia Institute of Technology
RUDRA KAPILA, Third Way
ROBERT KUMPF, Deloitte Consulting, LLP
JEFF LEE, Kronos Management, LLC
PAUL MAJSZTRIK, Solidia Technologies
THOMAS MALLOUK (NAS), University of Pennsylvania
MERCEDES MAROTO-VALER, Heriot-Watt University
PIOTR MONCARZ (NAE), E^{x}ponent
JOSÉ SANTIESTEBAN (NAE), ExxonMobil (retired)
SHUCHI TALATI, Carbon180
BRITTANY TARUFELLI, Pacific Northwest National Laboratory
GAVIN TOWLER (NAE), Honeywell
CATHY TWAY, Johnson Matthey

Although the reviewers listed above provided many constructive comments and suggestions, they were not asked to endorse the conclusions or recommendations of this report, nor did they see the final draft before its release. The review of this report was overseen by Andrew Brown, Jr. (NAE), Diamond Consulting and Delphi Automotive, and Christopher W. Jones (NAE), Georgia Institute of Technology. They were responsible for making certain that an independent examination of this report was carried out in accordance with the standards of the National Academies and that all review comments were carefully considered. Responsibility for the final content rests entirely with the authoring committee and the National Academies.

Contents

SUMMARY	1
1 INTRODUCTION AND SCOPE	**7**
1.1 Study Context	7
1.2 Overview of CO_2 Utilization Products, Infrastructure, and Societal Considerations	14
1.3 Study Statement of Task	17
1.4 References	18
2 EXISTING INFRASTRUCTURE FOR CO_2 UTILIZATION	**20**
2.1 CO_2 Sources	20
2.2 Existing Processes and Facilities Utilizing CO_2	26
2.3 Existing CO_2 Transport and Storage Infrastructure	28
2.4 Status of Enabling Infrastructure for CO_2 Utilization	32
2.5 Findings on Existing Infrastructure for CO_2 Utilization	38
2.6 References	39
3 POTENTIAL USES OF CO_2 IN COMMERCIAL PRODUCTS	**44**
3.1 Framing, Introduction, and Scope of Chapter	44
3.2 Future Sources of CO_2 for Utilization	50
3.3 Potential Utilization Products and Processes	51
3.4 Emerging, Pilot, and Commercial Facilities Utilizing CO_2	57
3.5 Priority Needs for CO_2-Derived Products That Could Contribute to a Net-Zero Carbon Future	57
3.6 Near-Term Opportunities, Synergies, and Needs	64
3.7 Findings and Recommendations on Potential Uses of CO_2 in Commercial Products	67
3.8 References	68
4 INFRASTRUCTURE CONSIDERATIONS FOR CO_2 UTILIZATION	**74**
4.1 CO_2 Capture	74
4.2 CO_2 Purification	78
4.3 CO_2 Transportation	82
4.4 CO_2 Conversion and Product Transportation	90
4.5 Enabling Infrastructure	92
4.6 Findings and Recommendations on Infrastructure Considerations for CO_2 Utilization	99
4.7 References	102

5	POLICY, REGULATORY, AND SOCIETAL CONSIDERATIONS FOR CO_2 UTILIZATION SYSTEMS	109
	5.1 Policy and Regulatory Considerations	109
	5.2 Current Regulatory Framework for Carbon Capture, Utilization, and Storage	113
	5.3 Societal Acceptance and Environmental Justice	119
	5.4 Findings and Recommendations for Policy, Regulatory, and Societal Considerations for CO_2 Utilization	124
	5.5 References	126
6	PRIORITY INFRASTRUCTURE OPPORTUNITIES FOR CO_2 UTILIZATION	129
	6.1 Infrastructure Funding and Investments	129
	6.2 Near-Term Versus Long-Term Infrastructure Strategies	131
	6.3 Findings and Recommendations on Priority Infrastructure Opportunities for CO_2 Utilization	133
	6.4 References	134

APPENDIXES

A	Committee Member Biographies	139
B	Disclosure of Conflicts of Interest	143
C	Information-Gathering Activities	145
D	Acronyms and Abbreviations	147

Summary

One essential component of maintaining a safe, stable climate is a global transition to a net-zero carbon dioxide (CO_2) emissions system, with no new net accumulation of CO_2 in the atmosphere. In 2019, CO_2 emissions in the United States totaled 5.26 gigatonnes (Gt), out of a total of 33.4 Gt CO_2 emissions globally. Among the highest priority actions to mitigate these emissions are decarbonizing the electric grid and transitioning energy end uses to decarbonized electricity, which will eliminate the majority of sources of emissions from fossil-fuel combustion. This long-term transition to net-zero emissions likely will eliminate the largest current uses of carbon-based products and processes, such as gasoline or diesel fuel for light-duty transportation and natural gas for heating in residential and commercial buildings. Mitigation of the remaining CO_2 emissions will occur via a number of different strategies, collectively termed carbon management. Large volumes of CO_2 will require long-term removal from the atmosphere, for instance, by storage in geologic formations underground or by sequestration in durable, long-lived products such as concrete, aggregates, some polymers, and nanostructured forms of carbon, such as carbon fiber. Short-lived carbon-based products, including fuels for some heavy-duty transportation applications, some polymers, and commodity chemicals, still will be needed and will have to be produced and used in a circular manner with no net emissions on a life cycle basis. Currently, the vast majority of embodied carbon in chemicals and materials (84 percent globally in 2020) is fossil derived, but in a circular carbon economy, this fossil carbon will need to be replaced with sustainable carbon from waste plastics and other carbon-based products, biomass-based carbon feedstocks, and CO_2. This report focuses on CO_2 utilization, in which CO_2 captured from flue gas streams, the atmosphere, or bodies of water is chemically transformed into a marketable product. The positioning of CO_2 utilization within the larger carbon management objective is illustrated in Figure S-1.

In the Energy Act of 2020, the U.S. Congress mandated that the U.S. Department of Energy (DOE) contract with the National Academies of Sciences, Engineering, and Medicine to analyze opportunities and challenges to advance CO_2 utilization technologies, develop the associated infrastructure, and establish markets for CO_2-derived products, considering a future in which carbon wastes participate in a circular carbon economy. For this first of two reports from the study, the committee was tasked with assessing the state of the infrastructure for CO_2 transportation, use, and storage as of early 2022 and identifying priority opportunities to improve and expand on that infrastructure to enable future CO_2 utilization development. The following paragraphs summarize the committee's approach to the study and its main findings, conclusions, and recommendations.

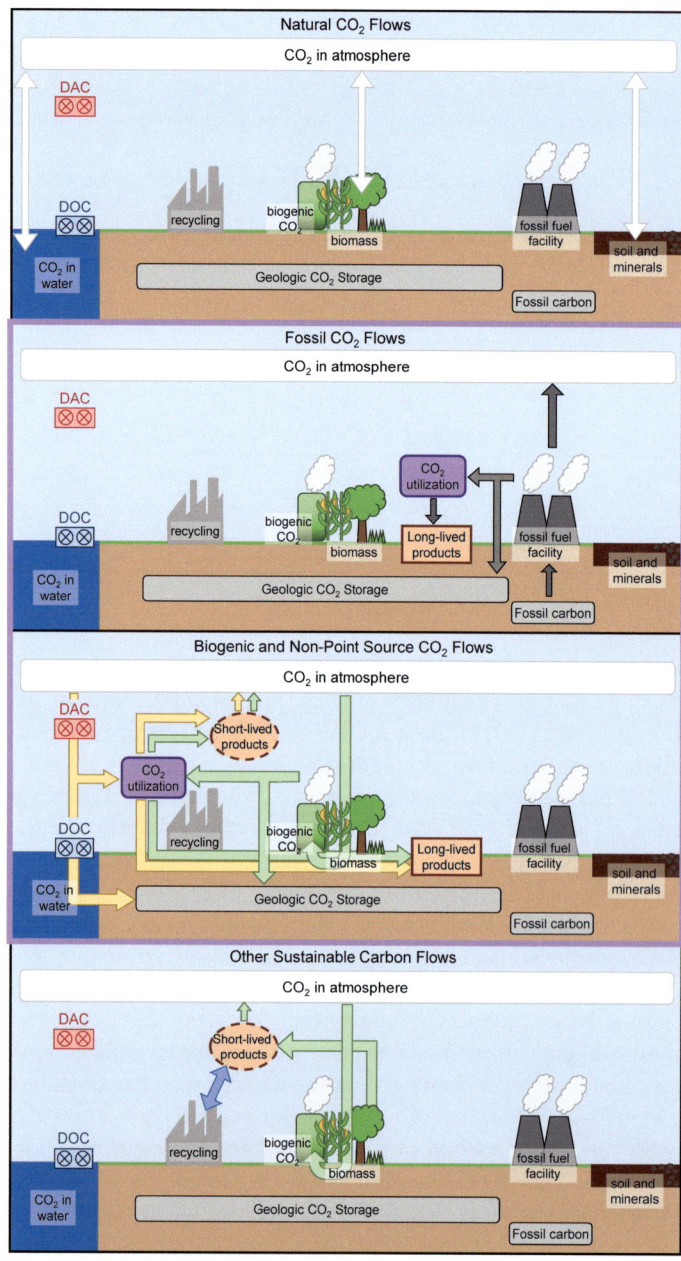

FIGURE S-1 Schematic of principal CO_2 and carbon flows in a circular carbon economy, with CO_2 utilization opportunities highlighted in the middle two panels (outlined in purple). The top panel shows natural flows of CO_2 into and out of the atmosphere, represented by white arrows. The second panel shows flows of fossil CO_2, represented by gray arrows, which in a net-zero case are minimized, with any remaining emissions from fossil facilities captured and diverted to geologic storage or utilized in long-lived carbon-based products. The third panel shows flows of biogenic CO_2 (green arrows) and CO_2 captured directly from the air or ocean (yellow arrows), which can be utilized sustainably in short- or long-lived carbon-based products or routed to geologic storage. The bottom panel shows other sustainable carbon flows, such as the conversion of biomass to short-lived products (green arrows) and recycling to make short-lived products (blue arrow). These schematics do not include all carbon flows that will exist in a circular carbon economy; for example, recycling could be an option for manufacturing long-lived as well as short-lived products.
NOTE: DAC = direct air capture; DOC = direct ocean capture.

The committee first identified existing infrastructure for CO_2 capture, transport, conversion, and geologic storage, as well as infrastructure for enablers of CO_2 utilization processes such as clean electricity, clean hydrogen, water, land use, and energy storage. The largest volume chemical utilization of CO_2 today is in urea synthesis; other products made on an industrial scale from CO_2 include organic carbonates, methanol, salicylic acid, and inorganic carbonates embedded in concrete. Pilot-scale efforts are under way to generate additional products from CO_2; however, the inherently higher energy requirements to use CO_2 as a feedstock in place of fossil carbon for hydrocarbon-based products and the lack of incentives to produce net-zero-emission products currently limit commercialization opportunities for CO_2 utilization. Furthermore, the committee found limited opportunities to leverage current CO_2 capture and transport infrastructure for future CO_2 utilization projects, since the majority of the existing infrastructure has been developed for enhanced oil recovery (EOR) and connects geologic or fossil-based sources of CO_2 with depleted oil reservoirs, routes that generally do not align with sustainable CO_2 utilization opportunities in a net-zero-emissions future. Nonetheless, the infrastructure expected to be developed for CO_2 capture and geologic storage (CCS) over the next decade also may be able to serve utilization projects. Therefore, the committee recommends that CCS developers design flexibility into their project plans to allow some of the CO_2 captured to be directed to utilization facilities (Recommendation 6.2). Given that many CO_2 utilization processes require significant amounts of clean electricity, clean hydrogen, and water, project developers also will need to consider the availability and accessibility of these inputs when designing CO_2 utilization infrastructure (Recommendations 4.4, 4.5, 4.6, and 6.3). DOE similarly should consider the availability and accessibility of geologic CO_2 storage, CO_2 transportation and product offtake infrastructure, clean electricity, and water when evaluating the design and siting of direct air capture and hydrogen hubs authorized in the Infrastructure Investment and Jobs Act (Recommendations 6.3 and 6.4).

To assess opportunities and challenges for future CO_2 utilization infrastructure development, the committee identified priority products that could be synthesized from CO_2 in a future circular carbon economy, separating these products into two classes, or "tracks," distinguished by their lifetime, and hence their ability to sequester carbon for long periods, as illustrated in Figure S-2. Track 1 products, with lifetimes greater than 100 years,[1] include concrete, aggregates, some polymers, and solid carbon materials that durably store CO_2. Track 2 products, with lifetimes less than 100 years, include chemicals, some polymers, and fuels that decay to CO_2 and re-emit it to the atmosphere on relatively short timescales, unable to store carbon for long periods. Importantly, these Track 1 and Track 2 products have different requirements for sustainable production consistent with net-zero-emissions goals. Track 1 products derived from non-fossil CO_2 sources, such as direct air capture or biogenic carbon, are net-negative-emissions compatible, while those derived from fossil CO_2 point sources are net-zero-emissions compatible. On the other hand, Track 2 products must be derived from non-fossil CO_2 sources to be net-zero-emissions compatible. In addition to considering the CO_2 source, assessments of net-negative- or net-zero-emissions compatibility require taking into account emissions associated with all other inputs (e.g., electricity, hydrogen, and heat), processes, product fates, and any co-products or waste generated. Such sustainability considerations influence the type and location of infrastructure, as discussed further below.

When developing infrastructure for CO_2 utilization, the entire value chain of CO_2 capture, purification, transportation, and conversion needs to be considered. Commercial technologies exist for purifying CO_2 to achieve the different requirements for capture, transport, utilization, and storage, but add cost to CO_2 utilization processes. Mineralization and biological CO_2 conversion are the most resilient to impurities of the CO_2 feedstock, and catalytic electro- and thermochemical CO_2 conversion are the most sensitive. Once captured (and purified, if required), CO_2 can be transported by a variety of means including pipeline, rail, truck, ship, and barge. In many cases, some combination of these transportation methods will be required. Multimodal transport infrastructure will be particularly beneficial for connecting small-scale emitters and capturers located far from main pipelines with larger industrial hubs containing utilization facilities, and for connecting large- or small-scale emitters and capturers with distributed small-scale users (e.g., concrete manufacturers). When accessible, CO_2 pipelines are the most cost-effective transportation option, as they can move the largest volumes of CO_2 and therefore benefit from

---

[1] The 100-year lifetime is chosen as the cutoff between Track 1 and Track 2 products to align with the United Nations Framework Convention on Climate Change.

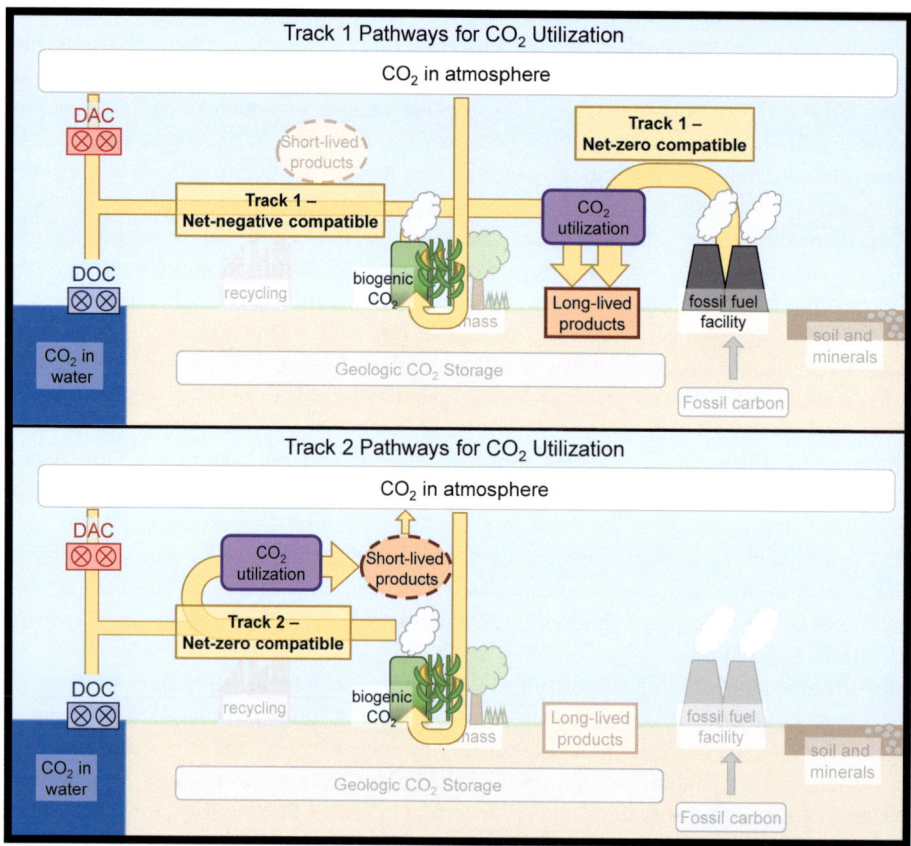

FIGURE S-2 Schematic of Track 1 (top) and Track 2 (bottom) pathways for CO_2 utilization. Track 1 pathways generate long-lived products and can be either net-zero compatible, if CO_2 is sourced from fossil emissions, or net-negative compatible if CO_2 is sourced from biogenic emissions or from direct air or water capture. Track 2 pathways generate short-lived products and can be net-zero compatible if CO_2 is sourced from biogenic emissions or from direct air or water capture. Track 1 pathways enable carbon sequestration in products, and Track 2 pathways enable circular carbon flows.
NOTE: DAC = direct air capture; DOC = direct ocean capture.

economies of scale. Of paramount importance in any transportation scenario is safety; the primary risks from CO_2 are physical damage due to rupture of pressurized equipment and asphyxiation. There is a considerable level of understanding regarding the risks associated with pipeline transportation of CO_2, both dense-phase and gaseous. One option under consideration for CO_2 transport is to repurpose natural gas pipelines, which could simplify acquisition of rights-of-way and potentially decrease capital costs. However, pipeline integrity and operational issues make this challenging. Each individual project would require rigorous techno-economic and cost-benefit analyses to determine its feasibility.

Infrastructure development decisions for CO_2 utilization depend not only on CO_2 capture, purification, and transportation requirements but also on the availability of enabling infrastructure for inputs such as clean electricity, clean hydrogen, and water and for transportation of the CO_2-derived product. Many CO_2 utilization processes require significant amounts of clean electricity and clean hydrogen. Increases in demand for zero-carbon-emissions electricity, including that required for CO_2 utilization, will influence transmission and distribution planning and load management. In the long term, post-2050, it is expected that the grid will need to accommodate all power demand with a reliable supply of net-zero-emissions electricity for all users, but in the short term, CO_2 utilization processes may require deployment of clean power generation and energy storage to enable the 24/7 operation likely required for economic viability. Given the complexities of transporting and storing hydrogen, the committee recommends that project designers co-locate facilities for hydrogen production and CO_2 utilization when feasible (Recommendation 4.5). Another required input for some CO_2 utilization processes is water. Though not expected to have a significant impact on water demand at a national scale, water requirements for CO_2 utilization could put stress on local water resources. Water will be an input to biological and chemical utilization processes, as a reaction medium, a reactant, or for process cooling. The committee therefore recommends siting CO_2 utilization facilities near a sustainable water source to support both short- and long-term needs and analyzing the effect of CO_2 utilization and the required enabling inputs (e.g., hydrogen) on local water demands, taking into consideration local environmental and justice impacts (Recommendation 4.6). Some CO_2 utilization processes, such as concrete production and biological CO_2 conversion to alcohols and hydrocarbons, are well suited for small-scale, distributed operations, with co-located CO_2 capture and conversion followed by transport of the CO_2-derived product. Optimizing the CO_2 transportation network to meet the needs of different utilization processes requires a systems-level analysis considering technological, environmental, economic, and societal factors (Recommendation 6.5).

In addition to developing technology and infrastructure, enabling and expanding CO_2 utilization requires consideration of policies, regulations, equity, and environmental justice. The most effective and efficient policies to enable CO_2 utilization would ensure that externalities from greenhouse gas emissions from all sources are captured in the full cost of using the technology and incentivize knowledge creation and reduce costs and risks for early adopters rather than subsidize specific technologies. In all cases, policies should support the CO_2 utilization industry toward societal goals, such as net-zero emissions, in a technology-agnostic manner. Policy makers should also take care that the policy regime does not create perverse incentives or excessively difficult regulatory environments (Recommendation 5.1). Establishing policies that overly incentivize CO_2 capture, for example, may create an unwanted incentive to produce emissions. Navigating the requirements to obtain permits from many different entities or managing the uncertainty in policy stability over the long term may decrease or deter investments in CO_2 utilization. The committee recommends that a single entity be responsible for guiding CO_2 utilization project developers through the permitting process, given the large number of organizations involved in granting permissions (Recommendation 5.4). Finally, the committee recognizes that widespread deployment of CO_2 utilization may have significant environmental, economic, and societal impacts. Total costs and benefits to society and their distribution should be assessed for each individual project using cost-benefit analyses. In particular, the equity and justice implications of CO_2 utilization projects should be examined and addressed. Furthermore, CO_2 utilization projects should involve early and sustained community engagement and address inequality, which may include terminating a project that does not receive local buy-in or for which regulators determine that the equity implications are unacceptable (Recommendation 5.6).

Given all of the above considerations, the committee identified two primary near-term opportunities for CO_2 utilization infrastructure investment: (1) using CO_2 sourced from bioethanol plants to make sustainable synthetic aviation fuel and (2) mineralization using fossil or non-fossil CO_2 sources to generate aggregates for construction materials, including concrete. In both cases, locating the CO_2 utilization operation in close proximity to enabling infrastructure (e.g., clean electricity and clean hydrogen) could prove cost-effective. While the committee deemed these two opportunities to be priorities in the near term given current knowledge and technologies and their potential scale of deployment, they are by no means the only promising CO_2 utilization opportunities. Consequently, and recognizing that infrastructure to capture and transport CO_2 for storage will likely be developed over the next decade, the committee recommends designing flexibility into that infrastructure such that it could be used to meet longer-term CO_2 utilization needs (Recommendation 6.2). One example would be developing industrial clusters that co-locate capture, utilization, and storage of CO_2 and incorporate relevant industries such as hydrogen production, chemical and fuel manufacturing, and low-carbon power generation. Specifically, the committee recommends that the Department of Energy consider co-locating at least one each of the hydrogen and direct air capture hubs authorized in the Infrastructure Investment and Jobs Act of 2021 (Recommendation 6.4). The best deployment and investment opportunities for CO_2 utilization should be identified using techno-economic, life cycle, and integrated systems analyses, while considering relevant regulatory and policy frameworks and factors that may influence societal acceptance via cost-benefit analyses (Recommendation 6.1). All infrastructure siting and development decisions should include best practices for substantive community engagement to ensure equitable outcomes.

Net-zero or net-negative CO_2 utilization can support the decarbonization transition by providing pathways for sustainable synthesis of carbon-based chemicals and materials in products, and carbon storage in durable, Track 1 products. However, few commercial examples of companies making net-zero or net-negative products via CO_2 utilization exist today, and for CO_2 utilization to meet its potential will require overcoming limitations and challenges related to technologies, processes, infrastructure, and business opportunities. These limitations and challenges include inherent inefficiencies that result in high costs; needs for discovery of and improvement in processes and technologies; requirements for large amounts of additional inputs, such as land, hydrogen, water, and electricity; elimination of the emissions associated with the entire life cycle of a CO_2 utilization process and product; limited existing CO_2 infrastructure; and societal concerns about infrastructure deployment, especially from environmental justice communities. Considering these challenges and opportunities, this report evaluates near-term needs for CO_2 utilization to play a role in a net-zero future.

1

Introduction and Scope

1.1 STUDY CONTEXT

As one essential component of maintaining a stable, safe climate, the world must transition to a net-zero, or even net-negative, carbon emissions[1] system, in which there is no net accumulation of carbon dioxide (CO_2) in the atmosphere. According to the Intergovernmental Panel on Climate Change 2022 report on climate change, global net anthropogenic greenhouse gas (GHG) emissions continued to rise and reached 59 ± 6.6 Gt CO_2e (gigatonnes of carbon dioxide equivalents) in 2019, although the rate of growth has fallen compared to the previous decade (IPCC 2022). Nonetheless, annual average emissions levels were higher between 2010 and 2019 than during any other time in human history. CO_2 emissions from fossil-fuel combustion and industry account for the largest percentage of GHG emissions (64 percent in 2019) and the largest absolute emissions growth since 1990 (67 percent). If global CO_2 emissions continue at current rates, models indicate that the world will exhaust the remaining carbon budget for limiting warming to 1.5°C before 2030, and before mid-century for limiting warming below 2°C (MCC 2022). Reducing reliance on fossil resources and mitigating GHG emissions would have co-benefits beyond limiting global warming, such as improving air quality by reducing criteria pollutants. The COVID-19 pandemic initially caused a global economic slowdown and corresponding decline in anthropogenic CO_2 emissions, most notably in the transportation sector. While most economies and emissions have already rebounded and even surpassed previous annual levels (IEA 2021), the long-term effect of the COVID-19 pandemic on anthropogenic emissions remains uncertain. Some behavioral changes from the COVID-19 pandemic, such as increased remote work and less business travel, could translate to a reduction in GHG emissions. Halting and ultimately reversing climate change will require more than behavior change, however. It will require net-zero or net-negative emissions over prolonged periods of time, fueled by large capital investments in innovative technologies.

Such a transition is challenging for many reasons, including that living things are carbon based, and carbon-containing manufactured products—such as fuels, building materials, plastics, and commodity chemicals—pervade the modern world. Currently, these products are derived primarily from fossil carbon, including coal, oil, and natural gas, and their production, use, and disposal typically result in accumulation of CO_2 in the atmosphere. Net-zero

---

[1] Net-zero carbon emissions is shorthand for net-zero greenhouse gas (GHG) emissions. The bulk of GHG emissions by volume is CO_2, with other GHG emissions often expressed in terms of equivalent amounts of carbon or CO_2. This study examines CO_2 utilization specifically, and thus, the committee will refer to net-zero CO_2 emissions and will discuss primarily volumes and masses of CO_2 emissions. Other GHGs are not considered in depth in this report, though they also must be addressed, in conjunction with CO_2 emissions, to reach a stable climate system. Throughout the report, net-zero CO_2 emissions are referred to as "net-zero."

systems do not require the entire elimination of carbon-based products and systems, or their CO_2 emissions, but they do require that flows of CO_2 to and from the atmosphere be balanced across total sources and sinks. Requiring net-zero CO_2 emissions will mean a significant reduction in many flows of carbon in the Earth system. For example, much gasoline and diesel fuel use for transportation will be eliminated in favor of clean electricity or hydrogen. For those uses of carbon-containing products that cannot be eliminated, the resulting CO_2 emissions will need to be balanced by diverting CO_2 from entering the atmosphere or removing CO_2 directly from the atmosphere.[2] Where needs for carbon-based materials remain, carbon wastes,[3] including CO_2 waste streams, can play a role as feedstocks for products. To put these needs into context, Box 1-1 highlights several studies examining the role of carbon capture, utilization, and storage (CCUS) in achieving net-zero emissions by 2050. This report focuses on CO_2 utilization, in which CO_2 feedstocks replace fossil-carbon feedstocks to help enable continued production, use, and disposal of carbon-based materials in a net-zero world with balanced flows of CO_2 to and from the atmosphere.[4] Sustainable CO_2 flows are illustrated in Figure 1-1.

BOX 1-1
Scenarios of Technology Needs for Net-Zero Emissions by 2050

Multiple studies, as well as local, state, and national plans, project a path to net-zero involving a shift in the energy system away from fossil fuels and toward a zero-emissions mix of energy carriers, such as hydrogen and electricity, and sources, such as renewable energy (e.g., Climate Mayors n.d.; DOS and EOP 2021; Larson et al. 2021; NASEM 2021; Paulos 2021; USCA 2019).[a] The role for carbon capture, utilization, and storage (CCUS) technologies in the future net-zero energy system is less certain. Implementation of CCUS may enable the persistence of more fossil-fuel combustion than would have remained without use of CCUS, along with its negative impacts. Alternatively, in a future without fossil-fuel combustion, CCUS could play a critical role in mitigating hard-to-abate, non-fossil-based CO_2 emissions, such as those released from limestone during cement production. Many studies of net-zero systems do envision a role for CCUS (NASEM 2021).

In November 2021, the United States put forward a Long-Term Strategy report to achieve net-zero emissions by no later than 2050 (DOS and EOP 2021). Achieving this net-zero goal will require reducing net U.S. emissions from roughly 6.6 Gt CO_2e in 2005 (and 4.7 Gt CO_2e in 2020) to zero by no later than 2050 (EPA 2022). The Long-Term Strategy report lays out a trajectory for the United States to achieve its goal, with contributions from all sectors of the economy, supportive regulations, and direct investments from government and corporations alike (DOS and EOP 2021). The report calls out five major categories for action: (1) decarbonizing electricity, (2) fuel switching, (3) energy efficiency, (4) reducing non-CO_2 emissions, and (5) carbon removal technologies. Figure 1-1-1 shows a representative pathway to net-zero on the left, but also demonstrates with the graphic on the right that there are multiple possible pathways to achieve this goal, which depend on the evolution of policies and technologies over time. Although the U.S. Long-Term Strategy does not explicitly call out CCUS technologies, the Infrastructure Investment and Jobs Act (IIJA) passed in November 2021 included more than $10 billion for CCUS projects (DOE 2021).

---

[2] Pathways for removing and sequestering CO_2 directly from the atmosphere, including direct air capture with sequestration, coastal carbon sequestration, and carbon uptake by forests and soils, are detailed in a 2019 National Academies report on *Negative Emissions Technologies and Reliable Sequestration* (NASEM 2019). Such negative emissions technologies are not the focus of this report. That National Academies report and others address the scale at which these solutions can be deployed and their projected effectiveness in meeting net-zero emissions targets (see, e.g., Carbon180 2022; EASAC 2018; EFI 2019; Mulligan et al. 2020).

[3] For the purposes of this report, carbon wastes are limited to CO_2 waste streams. A second report of the committee will also address coal-derived carbon wastes. Other non-CO_2 waste streams, including methane, biogas, plastic waste, used carbon-based products, and bio-based carbon feedstocks such as biomass and municipal, sanitary, and agricultural wastes, are outside the scope of the study.

[4] As further discussed in Box 1-2, CO_2 utilization for the purposes of this report is defined as a chemical transformation of CO_2 and therefore does not include enhanced oil recovery.

BOX 1-1 Continued

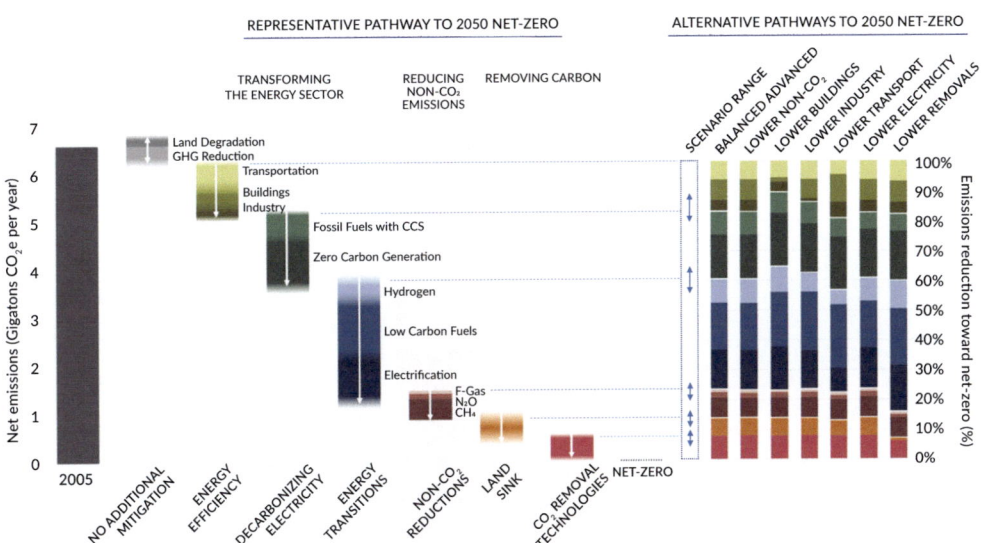

FIGURE 1-1-1 (Left) A representative pathway to net-zero GHG emissions in the United States by 2050, showing emissions reductions achieved via different actions: energy efficiency (chartreuse), decarbonizing electricity (green), energy transitions (blue), non-CO_2 reductions (brown), land sinks (orange), and CO_2 removal technologies (pink). **(Right)** Alternative pathways to net-zero emissions with different amounts of emissions reductions from each action, which are based on differences in policies and technologies over time.
SOURCE: U.S. Department of State, 2021, *The Long-Term Strategy of the United States: Pathways to Net-Zero Greenhouse Gas Emissions by 2050*, Washington, DC: DOS and the Executive Office of the President (EOP), https://whitehouse.gov/wp-content/uploads/2021/10/US-Long-Term-Strategy.pdf. Courtesy of DOS and EOP (2021).

For the electricity sector, the United States has set a goal to eliminate CO_2 emissions by 2035, in support of its 2050 net-zero target, which will likely be accomplished primarily with a shift to renewables and other emission-free electricity paired with energy storage. Retrofitting existing fossil power generation with carbon capture provides another pathway to clean, firm, dispatchable resources to enhance grid reliability, although the CCUS system requires additional energy. Carbon capture also could advance industrial decarbonization for those processes that cannot be powered with clean electricity or decarbonized heat, or that produce CO_2 as a by-product of reaction. Clean hydrogen production may be enabled by deploying carbon capture technologies on existing steam methane reforming (SMR) assets, such as those used in the refining and petrochemicals industry and for making hydrogen for ammonia production, that have multiple decades of remaining useful life. (See Abramson et al. [2022] for locations of existing hydrogen and ammonia production in the United States, the majority of which involve SMR.) New guidelines issued in spring 2022 from the White House Council on Environmental Quality (CEQ) Report to Congress on Carbon Capture, Utilization, and Sequestration found that responsible CCUS deployment is also critical to enable CO_2 removal from ambient air (CEQ 2021).

In addition to the national plans and goals publicly stated by the U.S. government, many published scenarios and models describe pathways to address climate change through U.S. and global actions to reduce CO_2 and other GHG emissions and to remove carbon, including through CCS. A McKinsey analysis provides one possible emissions reduction scenario for the United States to remain on a pathway that limits global temperature rise to 1.5 degrees Celsius and forecasts annual investment of $35 billion by 2025 in carbon capture equipment and CO_2 transportation infrastructure to achieve annual abatement of 230 million metric tons of CO_2e by 2030 (Clune et al. 2022). These investments reflect capital expenditures only and do not include costs for operating the equipment and infrastructure once deployed. When compared

continued

BOX 1-1 Continued

against shifting to renewable electricity or electrification of transportation and industry, the McKinsey analysis estimates that CCS deployment, specifically on the Gulf Coast and in the Midwest, would have a higher cost per tonne of abated CO_2e (Clune et al. 2022).

Princeton University's Net-Zero America Project (Larson et al. 2021) models deployment of CCUS at a large scale in all scenarios and considers it a key pillar for decarbonization. CCUS and its requisite infrastructure are implemented in cement production, gas- and biomass-fired power generation, natural gas reforming, biomass-derived fuels, as well as direct air capture (DAC). All scenarios but one implement CCS, and in those scenarios, CO_2 is captured from over 1,000 facilities and geologically sequestered on the order of 0.9 to 1.7 Gt CO_2 per year. The cumulative range of investment from 2021 to 2050 (in 2018 dollars) varies from $167 billion to $225 billion for CO_2 pipelines and from $38 billion to $60 billion for storage wells, facilities, and construction costs. The other scenario that does not consider underground CO_2 storage captures and utilizes 0.7 Gt CO_2 per year for the production of synthetic liquids.

The Net Zero by 2050 Scenario (NZE) from the International Energy Association's (IEA's) World Energy Outlook 2021 presents a pathway for the global energy sector to achieve net-zero CO_2 emissions by 2050 and provides an estimate of the level of investment required globally (IEA 2021). In the NZE, levels of carbon capture and removal rise marginally over the next 5 years, but by 2050 account for a total of 7.6 Gt CO_2, of which almost 50 percent is from fossil-fuel combustion, 20 percent is from industrial processes, and around 30 percent is from bioenergy use with CO_2 capture and DAC. This radical transformation would require global investments in CCUS of $160 billion per year by 2050. The IEA estimates that if no new fossil-fuel CCUS projects are developed beyond those already under construction or approved, the additional investments in other technologies required to reach net-zero emissions in 2050 would be around $15 trillion, more than 10 percent higher than the required investments in the NZE Scenario (IEA 2021).

[a] All energy sources and carriers currently have some associated GHG emissions on a life cycle basis, and these must be offset by GHG-negative activities to yield an overall net-zero-emission system.

CO_2 utilization may be an important part of a future net-zero world, but there are many challenges to its use at large scale today. For the CO_2 utilization opportunities within the scope of this study (i.e., chemical transformations of CO_2, see Box 1-2 for details), few commercial processes exist as of 2022, and those that exist are performed at limited scale. There are many challenges to commercializing CO_2 utilization processes, including (1) the inherent difficulty, given the thermodynamic stability of CO_2, of transforming it into desired products, relative to using fossil-carbon feedstocks to synthesize the same product; (2) the lack of infrastructure to enable CO_2 utilization at a large scale for purposes other than enhanced oil recovery (EOR); (3) the need for inputs of clean (i.e., low- or zero-carbon)[5] hydrogen, electricity, and heat; and (4) the resulting higher cost of CO_2-derived products compared to alternative products. Discovering, developing, and commercializing CO_2 utilization processes and products are important activities for the nation and the world as it transitions to a net-zero economy. In response to this need, the Consolidated Appropriations Act of 2021, including the Energy Act of 2020, legislated a mandate for the U.S. Department of Energy (DOE) to enter into an agreement with the National Academies to conduct a study "to assess any barriers and opportunities relating to commercializing carbon, coal-derived carbon,[6] and carbon dioxide in the United States" (Consolidated Appropriations Act 2021). The mandate further required study of CO_2 utilization markets, infrastructure, and research, development, and deployment. The study statement of task is reproduced at the end of this chapter.

---

[5] "Clean" or low- or zero-carbon refers to the CO_2 emissions associated with the input (e.g., hydrogen, electricity, or heat), which depend on the methods by which it is produced. The committee follows DOE's definition and therefore defines clean hydrogen as having a GHG footprint of less than 2 kg CO_2e per kg H_2.

[6] Coal-derived carbon will be examined in the second report of the study.

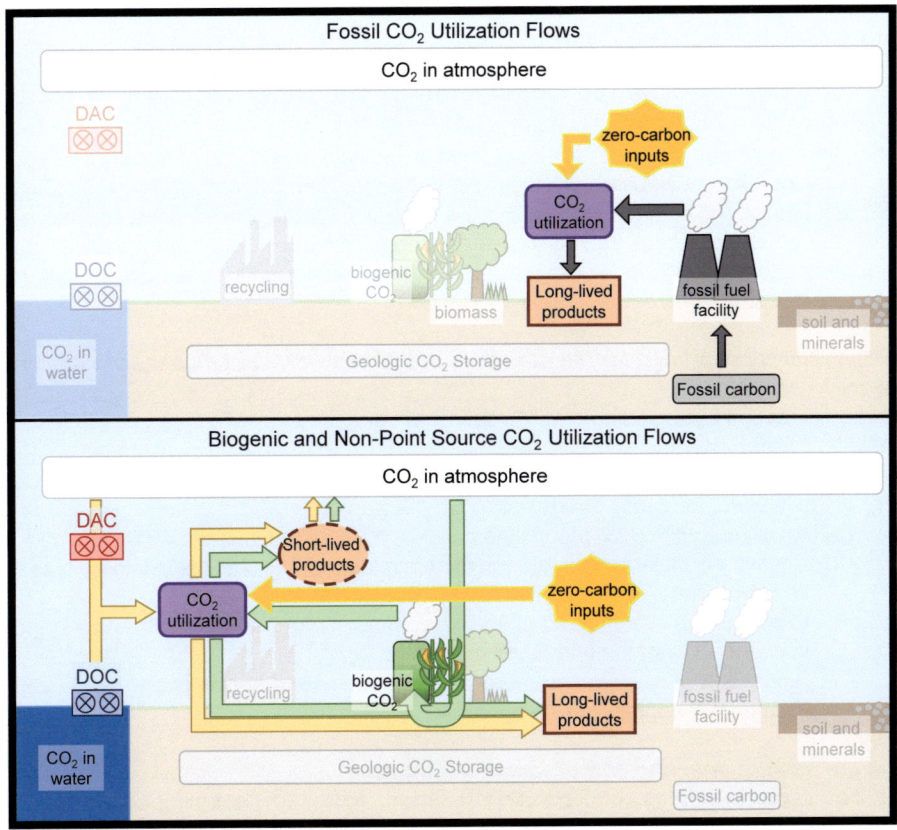

FIGURE 1-1 Sustainable CO_2 utilization flows, showing options for fossil-derived CO_2 (top) and CO_2 from biogenic emissions or from direct air or water capture (bottom). Fossil-derived CO_2 can be utilized to make long-lived carbon-based products (gray arrows), while CO_2 from biogenic sources (green arrows) and from direct air or water capture (light-yellow arrows) can be used to make short- or long-lived carbon-based products. For CO_2 utilization processes to have low carbon emissions on a life cycle basis, other energy and feedstock inputs must be net-zero carbon, as illustrated with the orange star and arrow in both panels. NOTE: DAC = direct air capture; DOC = direct ocean capture.

Per its statement of task, the study will produce two reports, with this first report focusing on the current state of infrastructure for CO_2 transportation, use, and storage in the United States, as well as priority opportunities and challenges to develop that infrastructure to enable future CO_2 utilization processes and markets in a safe, cost-effective, and environmentally benign manner. The second report will evaluate in more detail the potential markets for products derived from CO_2 and coal waste; the economic, environmental, and climate impacts of CO_2 utilization infrastructure; and research, development, and demonstration needs to enable commercialization of CO_2 utilization technologies and processes. In addressing the first report's tasking, the committee first considered which products and processes are required for, or could contribute to, a future net-zero economy. Then, for those products and processes, the committee evaluated the associated infrastructure requirements, including for CO_2 and/or product transportation, CO_2 conversion, and other necessary inputs such as clean hydrogen and clean electricity. (See Box 1-2 for more detail about the definition and scope of CO_2 utilization used in this study.) The committee considered opportunities and challenges for the policy, regulatory, and societal aspects of developing CO_2 utilization systems. Key topics addressed in the report are introduced below.

BOX 1-2
Scope Considered in the Study

Carbon utilization carries different definitions, and the system scope and boundary for aspects of carbon utilization vary for different industries, disciplines, and applications. In considering the study's mandate and its statement of task, the committee identified the limits of its scope and defined several aspects of its work. It identified certain processes as in or out of the scope of carbon utilization within its consideration. It also developed certain definitions for the purposes of responding to its statement of task, including of CO_2 sources and infrastructure.

How is carbon utilization defined for the purposes of this report? What classes of carbon utilization processes are in scope?
- Carbon utilization is defined here as the chemical transformation of CO_2 into a product with market value.
  - The committee is only considering carbon dioxide utilization, and will use that terminology going forward, generally shortened to CO_2 utilization.
  - Carbon dioxide utilization processes considered in scope include:
    - Chemical transformation through mineralization of CO_2 into materials such as aggregates and concrete;
    - Production of organic and inorganic commodity and specialty chemicals and materials in chemical systems and facilities; and
    - Production of organic and inorganic commodity and specialty chemicals and materials in biological engineered systems and facilities.

What terms are used to describe carbon dioxide utilization?
- **Carbon capture** is the act of separating, purifying, and/or concentrating CO_2, from an industrial or other gas stream, from the ambient air, or from bodies of water. Carbon capture from point sources prevents the release of CO_2 into the atmosphere, whereas carbon capture from ambient air or bodies of water removes CO_2 from the atmosphere.
- **Carbon utilization, also called carbon dioxide utilization,** as defined above is the process of transforming CO_2 into a marketable product.
- **Carbon capture, utilization, and storage (CCUS)** generally is not used in this report because it conflates three rather different processes. Instead, carbon capture, or utilization, or storage/sequestration is described individually, or in relevant combinations. CCUS is used as a term when it is used in statutes or regulations, or when it is appropriate to describe all three aspects of carbon management (Olfe-Kräutlein et al. 2022).
- **Carbon removal** describes a system that captures CO_2 from the atmosphere or bodies of water and durably stores it. This could involve CO_2 utilization processes depending on the CO_2 source type, means of storage, and storage lifetime.

What processes and products of carbon utilization are considered out of scope for the purposes of this study?
- Carbon utilization processes that begin with non-CO_2 waste streams, such as methane, biomass materials, and municipal solid wastes. Utilization of coal waste streams is not considered carbon utilization in this report, though these waste streams will be addressed in their own right in the committee's second report.

- Uses of CO_2 that do not involve a chemical transformation, such as in enhanced oil recovery (EOR), fire suppression, or beverage carbonation.
- Chemical transformations of CO_2 resulting in nontraded products and goods, such as increased soil carbon and ocean mineralization.
- Production of terrestrial biomass from CO_2, such as grasslands, forests, and field crops.

What processes and feedstocks are out of scope but will be discussed in a limited fashion as they interact with and impact carbon utilization systems and markets?
- Emissions pathways that are exclusively negative emissions aiming to store CO_2 for periods of hundreds or thousands of years or more (see Box 1-3 for description of carbon storage lifetimes), including long-term geologic storage of CO_2.
- Non-CO_2, bio-based carbon feedstocks such as biomass and municipal, sanitary, and agricultural wastes.

How are the emissions from all aspects of carbon dioxide utilization processes accounted for? Are processes that do not permanently sequester CO_2 in scope?
- This report considers the full life cycle of CO_2 utilization processes when accounting for net CO_2 emissions.
- CO_2 utilization processes that durably sequester CO_2 in a product are in scope, as are processes that contribute to circular flows of CO_2 from a source, through use in a product, to disposal or generation of CO_2, as long as the CO_2 eventually emitted could be reused.

What CO_2 sources and forms are under consideration as carbon waste feedstocks for utilization?
- CO_2 captured from the atmosphere through direct air capture.
- CO_2 captured from point sources before emission to the atmosphere, such as from power plants or industrial facilities.
- CO_2 dissolved in natural or other bodies of water, where those waters are used as feedstocks or reaction media for carbon utilization processes.

What is infrastructure, for the purposes of this study?
- Infrastructure to capture CO_2 (from the sources defined above) for utilization.
- Transportation of CO_2 feedstocks and products (pipelines, other freight systems).
- Production facilities (including pilot plants).
- Infrastructure for production and transport of other required inputs and utilities (e.g., hydrogen, clean electricity, water, energy storage).
- CO_2 storage for utilization processes.
- Integrated infrastructure for CO_2 storage and utilization.

What infrastructure is out of scope for the study, but will be discussed as it relates to and impacts the infrastructure for CO_2 utilization?
- Infrastructure for geologic carbon sequestration and for carbon capture that is *only* destined for subsurface storage or EOR.
- Institutional infrastructure (e.g., government organizations and financial institutions).
- Human infrastructure (e.g., education systems, workforce training, and employment opportunities).

1.2 OVERVIEW OF CO_2 UTILIZATION PRODUCTS, INFRASTRUCTURE, AND SOCIETAL CONSIDERATIONS

The following sections provide background on the potential future demand for CO_2 utilization products in a net-zero system; infrastructure for CO_2 capture, transport, utilization, and storage; and current policies, regulations, and economic and societal considerations for CO_2 utilization. Box 1-3 examines the role of life cycle assessment and product lifetime in assessing the CO_2 emissions impact of a CO_2 utilization product.

1.2.1 Demand for CO_2 Utilization Products in Net-Zero Systems

CO_2 utilization will perform two key roles in future net-zero systems: serving as a feedstock for manufacturing carbon-based products sustainably, as well as storing carbon in solid form in durable products, sequestering it from the atmosphere for climate-relevant timescales. Key product categories, examined in Chapter 3, include construction materials (concrete and aggregates), chemicals and fuels (single- and multi-carbon molecules), polymers and polymer precursors, elemental carbon materials, and niche products. These processes require sources of CO_2, which can be collected from a diverse array of emitting facilities with various concentrations, impurities, and CO_2 volumes, or from the vast but low-concentration resources in the atmosphere and bodies of water. Further explored in Box 1-3 and Figure 1-3-1, long-lived CO_2-based products (so-called Track 1 products, with carbon storage of >100 years) offer opportunities for durable CO_2 storage, while shorter-lived products (Track 2, carbon storage of <100 years) can be enablers of a circular carbon economy to ensure continued access to essential carbon-containing products (Sick et al. 2022). The production and use of these two respective product types therefore will have different climate impact for each CO_2 molecule used. CO_2 utilization complements alternative sustainable sources of carbon, including product recycling and biomass, as well as alternative ways to meet societal needs, for example, using electricity or hydrogen as energy carriers in lieu of hydrocarbon fuels. Rigorous and transparent life cycle assessments that follow harmonized approaches provide the basis for decision making and reporting at every stage of research, development, and deployment.

1.2.2 Infrastructure for CO_2 Capture, Transport, Use, and Storage

Infrastructure to enable production and use of carbon-based products in a circular carbon economy includes technologies for CO_2 capture, methods to transport CO_2 and/or CO_2-derived products, facilities and technologies to transform CO_2 into useful products, and reservoirs for geologic CO_2 storage. CO_2 can be captured from point sources such as power plants and industrial facilities, from the atmosphere through direct air capture (DAC), or from the ocean and other bodies of water containing dissolved CO_2. Chapter 2 provides details about potential sources of CO_2 for utilization in the United States. Once the CO_2 has been captured, it can be chemically transformed on-site into a valuable product, transported for utilization elsewhere, or transported for long-term geologic storage. The primary mode of transportation for CO_2 is via pipeline, which can accommodate large volumes and involves transporting CO_2 primarily in its supercritical fluid state. Other transportation methods, which typically transport CO_2 in its liquid state, include truck, rail, ship, and barge. Chapter 2 provides more information about the status of CO_2 transport by each of these methods in the United States. As detailed in Chapter 4, in some cases, a combination of transport modes could be the optimal solution to move CO_2 from the source to the point of storage or use; in other cases, it may be desirable to utilize CO_2 at the point of capture and transport the CO_2-derived product instead.

As discussed in Chapter 2, several current CO_2 utilization processes transform CO_2 into a useful product, the largest of which is the synthesis of urea. Other smaller-scale commercial processes include synthesis of salicylic acid, methanol, and organic carbonates. Emerging facilities that co-locate sustainable CO_2 capture and utilization are currently in operation in the United States or worldwide (see, e.g., Circular Carbon Network 2022). For production of chemicals, examples include Carbon Recycling International's renewable methanol plant in Iceland, which has been operating since 2012 and produces 4,000 tons of methanol per year from CO_2, and two facilities in China that have partnered with LanzaTech on biological CO_2 conversion to commodity chemicals (CRI n.d.;

BOX 1-3
CO₂ Utilization Product Lifetimes and Emissions Abatement

CO_2 utilization, as defined in this report, focuses on producing commercial products from CO_2 in a future net-zero emissions economy. The act of using CO_2 as a carbon feedstock does not necessarily result in net-zero emissions for a product. All assessments of the net-positive, net-zero, or net-negative emissions status of CO_2 utilization products have to estimate the full product life cycle (Tanzer and Ramírez 2019). Aspects of the product life cycle need to be estimated comprehensively and included in the emissions balance, including any associated upstream or downstream greenhouse gas emissions.

The lifetime of carbon in CO_2 utilization products ranges from hundreds or thousands of years to mere days. Long-lived products, which hold carbon on geologic timescales of hundreds to thousands of years, may in fact be designed to store carbon in a manner intended to be permanent, such as CO_2 used to produce concrete and aggregates, or other solid carbon products, including some durable plastics and polymers. Products with medium lifetimes of tens to 100 years, such as some plastics and polymers, and short lifetimes of days to months, such as commodity chemicals and fuels, do not result in durable, climate-effective carbon removal. The different possible end-of-life outcomes for these products also need to be considered in evaluating the emissions impact of the CO_2 utilization process.

The carbon removal outcome of products that also provide long-term, durable storage differs from that of those with only medium- or short-term storage capability. Notably, if products with long-term storage capability are produced with emissions drawn from the atmosphere, then they may result in carbon removal; if they are produced with emissions from fossil-fuel combustion, then they may still be net-zero emissions technologies, depending on emissions required to produce the material itself and its inputs. Products with short-term storage such as some chemicals and fuels can at best be net-zero emissions if their carbon is sourced from the atmosphere rather than from fossil-fuel emissions. Short-lived products in some cases may have net-positive emissions if their carbon is sourced from fossil fuels, and thus do not qualify as net-zero emission CO_2 utilization (see Figure 1-3-1). Estimating the greenhouse gas implications of different CO_2 utilization processes requires using detailed life cycle assessment.

	Track 1: Long duration of durable carbon in product (lifetime >100 years)	Track 2: Short duration of nondurable carbon in product (lifetime <100 years)
CO_2 from biogenic sources or captured from the atmosphere or bodies of water - and - utilization processes and other inputs are net-zero or net-negative carbon	Negative emissions or carbon removal capable	Net-zero emissions capable
CO_2 from point-source fossil-fuel combustion - or - utilization processes and other inputs are net-positive carbon	Net-zero emissions capable	Net-positive emissions

FIGURE 1-3-1 Net greenhouse gas emissions implications of carbon source, carbon storage duration in product, and process emissions. Processes are capable of achieving net-zero emissions or negative emissions if they result in long-duration carbon storage in the product and/or if their carbon is sourced from biogenic emissions or captured from air or water. Short-duration products with carbon sourced from fossil fuels will have net positive emissions. Estimation and assignment of emissions to a given system or process requires a full life cycle assessment.
NOTE: For a Track 1 product utilizing fossil CO_2 to be classified as net-zero emissions, it needs to store more carbon in the product than was released in its formation and the other processes involved in its life cycle.

McCoy 2022). Both Carbon Recycling International and LanzaTech have additional projects under development or construction worldwide. For mineralization examples, there are a growing number of companies that provide CO_2-cured ready-mix concrete for commercial projects (CarbonCure 2022). The 2021 IIJA authorized funding across the CO_2 capture, transportation, use, and storage value chain, including $2.1 billion for CO_2 Transportation Infrastructure Finance and Innovation (§ 40304), $310 million for the Carbon Utilization Program (§ 40302), and $3.5 billion for establishing four regional DAC hubs (§ 40308) (IIJA 2021).

Although long-term geologic storage of CO_2 is not a focus of this report, it is still critical to consider when planning infrastructure development for CO_2 utilization, as explained in Chapter 4. The main geological formations being considered for CO_2 storage are saline reservoirs, depleted oil and gas reservoirs, and unmineable coal seams (Jones and Lawson 2021). Figure 1-2 illustrates the locations of these geologic reservoirs in relation to CO_2 pipelines, power plants, and other point-source emitters. Within the context of the existing infrastructure for CCUS and other enabling industries (e.g., hydrogen, clean electricity, and water), Chapters 4 and 6 present considerations and opportunities for further CO_2 utilization infrastructure development.

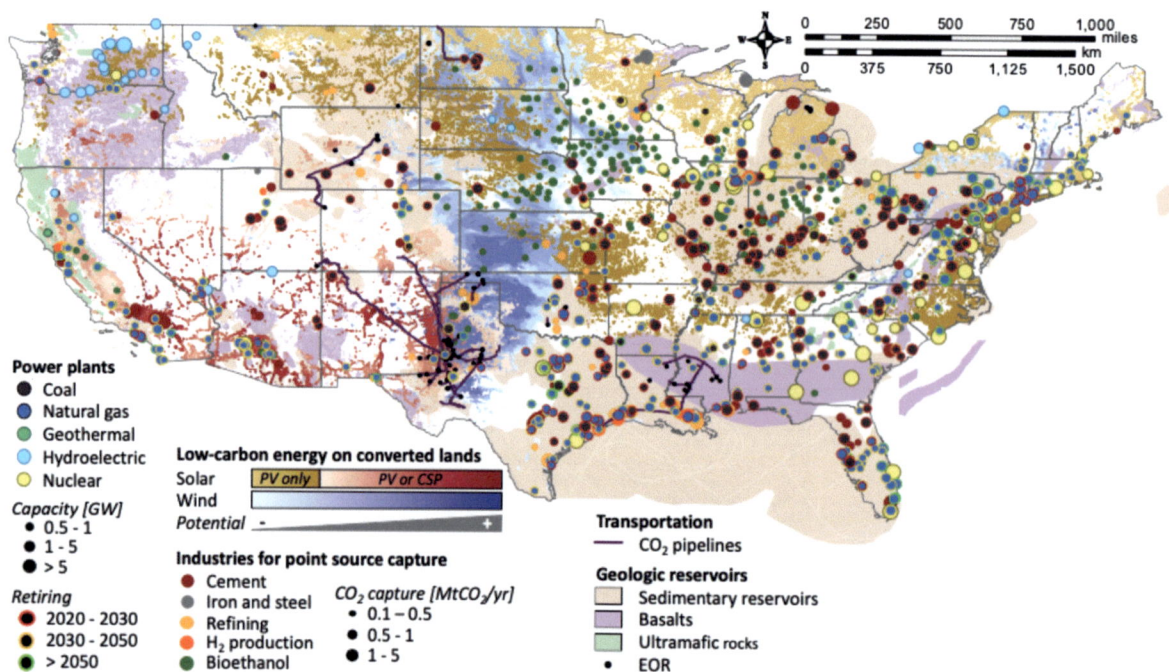

FIGURE 1-2 Power plant and industrial sources, pipelines, and geologic reservoirs for CO_2 in the contiguous United States, along with renewable solar and wind power opportunities on converted lands.
SOURCE: Adapted from National Academies of Sciences, Engineering, and Medicine, 2021, *Accelerating Decarbonization of the U.S. Energy System,* Washington, DC: The National Academies Press, https://doi.org/10.17226/25932.

1.2.3 Societal Considerations, Including Environmental Justice, Public Acceptance, and Policy and Economic Issues

In addition to the scientific and engineering considerations for developing technologies and infrastructure, various socioeconomic factors must be taken into account for successful deployment of CO_2 utilization. Economic, policy, regulatory, and community engagement tools exist that, when strategically implemented with incorporation of environmental justice principles, can facilitate both individual projects and the development of a widespread CO_2 utilization industry. Certain incentives can enable emerging technologies to compete with incumbent products and processes and encourage knowledge spillover, that is, the creation and sharing of information and best practices, to accelerate growth of this nascent industry. At an individual project level, navigating the regulatory framework and permitting processes, in addition to securing societal acceptance, is crucial for project success. Meaningful community engagement is important to ensure equitable distribution of the harms, benefits, and risks of these projects as these technologies are implemented for a more sustainable future. Chapter 5 expands on these socioeconomic tools.

1.3 STUDY STATEMENT OF TASK

The report that follows is the first of two reports produced by the Committee on Carbon Utilization Infrastructure, Markets, Research and Development. This first report focuses on the current state of infrastructure for CO_2 transportation, use, and storage, and priority opportunities to develop, improve, and expand that infrastructure to enable carbon utilization. The full study statement of task is as follows:

The National Academies of Sciences, Engineering, and Medicine will convene an ad hoc committee to assess infrastructure and research and development needs for carbon utilization, focused on a future where carbon wastes are fundamental participants in a circular carbon economy. In particular, the study will focus on regional and national market opportunities, infrastructure needs, and the research and development needs for technologies that can transform carbon dioxide and coal waste streams into products that will contribute to a future with zero net carbon emissions to the atmosphere. The committee will analyze challenges in expanding infrastructure, mitigating environmental impacts, accessing capital, overcoming technical hurdles, and addressing geographic, community, and equity issues for carbon utilization.

The committee will provide a first report which:
1) Assesses the state of infrastructure for carbon dioxide transportation, use, and storage as of the date of the study; including pipelines, freight transportation, electric transmission, and commercial manufacturing facilities.
2) Identifies priority opportunities for development, improvement, and expansion of infrastructure to enable future carbon utilization opportunities and market penetration. Such priority opportunities will consider how needs for carbon utilization infrastructure will interact with and capitalize on infrastructure developed for carbon capture and sequestration.

The committee will develop a second report that will evaluate the following:
1) Markets
   a. Identify potential markets, industries, or sectors that may benefit from greater access to commercial carbon dioxide to develop products which may contribute to a net zero carbon future; identify the markets that are addressable with existing utilization technology, and which still require research, development, and demonstration;
   b. Determine the feasibility of, and opportunities for, the commercialization of coal-waste-derived carbon products in commercial, industrial, defense, and agricultural settings; for medical, construction, and energy applications; and for the production of critical minerals;
   c. Identify appropriate federal agencies with capabilities to support small business entities; and determine what assistance those federal agencies could provide to small business entities to further the development and commercial deployment of carbon-dioxide based products.

2) Infrastructure
   a. Building off the study's first report, assess infrastructure updates needed to enable safe and reliable carbon dioxide transportation, use, and storage for carbon utilization purposes. Assessment of infrastructure will consider how carbon utilization fits into larger carbon capture and sequestration infrastructure needs and opportunities;
   b. Describe the economic, climate, and environmental impacts of any well-integrated national carbon dioxide pipeline system as applied for carbon utilization purposes, including suggestions for policies that could: (i) improve the economic impact of the system; and (ii) mitigate climate and environmental impacts of the system.
3) Research, Development, and Demonstration
   a. Identify and assess the progress of emerging technologies and approaches for carbon utilization that may play an important role in a circular carbon economy, as relevant to markets determined in section 1a.
   b. Assess research efforts under way to address barriers to commercialization of carbon utilization technology, including basic, applied, engineering, and computational research efforts, and identify gaps in the research efforts;
   c. Update the 2019 National Academies comprehensive research agenda on needs and opportunities for carbon utilization technology RD&D, focusing on needs and opportunities important to commercializing products that may contribute to a net zero carbon future.

The first and second reports will provide guidance to infrastructure funders, planners, and developers and to research sponsors, as well as research communities in academia and industry, regarding key challenges needed to advance the infrastructure, market, science, and engineering required to enable carbon utilization relevant for a circular carbon economy.

1.4 REFERENCES

Abramson, E., E. Thomley, and D. McFarlane. 2022. *An Atlas of Carbon and Hydrogen Hubs for United States Decarbonization*. Minneapolis, MN: Great Plains Institute. https://scripts.betterenergy.org/CarbonCaptureReady/GPI_Carbon_and_Hydrogen_Hubs_Atlas.pdf.

Carbon180. 2022. "Deep Dives." https://carbon180.org/deep-dives.

CarbonCure. 2022. "Projects." https://www.carboncure.com/projects.

CEQ (Council on Environmental Quality). 2021. *Council on Environmental Quality Report to Congress on Carbon Capture, Utilization, and Sequestration*. Washington, DC: Executive Office of the President. https://www.whitehouse.gov/wp-content/uploads/2021/06/CEQ-CCUS-Permitting-Report.pdf.

Circular Carbon Network. 2022. "Innovator Index." https://circularcarbon.org/innovator-index.

Climate Mayors. n.d. "Cities Climate Action Compendium." https://climatemayors.org/wp-content/uploads/2020/12/Cities_Climate_Action_Compendium_180105-1.pdf.

Clune, R., L. Corb, W. Glazener, K. Henderson, D. Pinner, and D. Walter. 2022. "Navigating America's Net-Zero Frontier: A Guide for Business Leaders." McKinsey & Company. https://www.mckinsey.com/capabilities/sustainability/our-insights/navigating-americas-net-zero-frontier-a-guide-for-business-leaders.

Consolidated Appropriations Act. 2021. "Division Z—Energy Act of 2020, § 969A." In H.R.133 - Consolidated Appropriations Act, 2021. Public Law 116-260. 116th Congress (2019–2020). https://science.house.gov/imo/media/doc/Energy%20Act%20of%202020.pdf.

CRI (Carbon Recycling International). n.d. "Projects." https://www.carbonrecycling.is/projects.

DOE (U.S. Department of Energy). 2021. "DOE Fact Sheet: The Bipartisan Infrastructure Deal Will Deliver for American Workers, Families and Usher in the Clean Energy Future." https://www.energy.gov/articles/doe-fact-sheet-bipartisan-infrastructure-deal-will-deliver-american-workers-families-and-0.

DOS and EOP (U.S. Department of State and Executive Office of the President). 2021. *The Long-Term Strategy of the United States: Pathways to Net-Zero Greenhouse Gas Emissions by 2050*. https://whitehouse.gov/wp-content/uploads/2021/10/US-Long-Term-Strategy.pdf.

EASAC (European Academies Science Advisory Council). 2018. *Negative Emission Technologies: What Role in Meeting Paris Agreement Targets?* EASAC Policy Report 35. Germany: European Academies Science Advisory Council. https://unfccc.int/sites/default/files/resource/28_EASAC%20Report%20on%20Negative%20Emission%20Technologies.pdf.

EFI (Energy Futures Initiative). 2019. *Clearing the Air: A Federal RD&D Initiative and Management Plan for Carbon Dioxide Removal Technologies – Summary Report*. Washington, DC. https://static1.squarespace.com/static/58ec123cb3db2bd94e057628/t/5d899dcd22a4747095bc04d5/1569299950841/EFI+Clearing+the+Air+Summary.pdf.

EPA (U.S. Environmental Protection Agency). 2022. "Inventory of U.S. Greenhouse Gas Emissions and Sinks: 1990–2020." EPA 430-R-22-003. https://www.epa.gov/ghgemissions/draft-inventory-us-greenhouse-gas-emissions-and-sinks-1990-2020.

IEA (International Energy Agency). 2021. *Net Zero by 2050: A Roadmap for the Global Energy Sector*. Paris. https://iea.blob.core.windows.net/assets/deebef5d-0c34-4539-9d0c-10b13d840027/NetZeroby2050-ARoadmapfortheGlobalEnergySector_CORR.pdf.

IIJA (Infrastructure Investment and Jobs Act). 2021. H.R.3684 - Infrastructure Investment and Jobs Act. Public Law 117-58. 117th Congress (2021–2022). https://www.congress.gov/bill/117th-congress/house-bill/3684/text.

IPCC (Intergovernmental Panel on Climate Change). 2022. "Summary for Policymakers," H.-O. Pörtner, D.C. Roberts, E.S. Poloczanska, K. Mintenbeck, M. Tignor, A. Alegría, M. Craig, et al., eds. In *Climate Change 2022: Impacts, Adaptation, and Vulnerability*. Working Group II Contribution to the IPCC Sixth Assessment Report, H.-O. Pörtner, D.C. Roberts, S. Langsdorf, S. Löschke, V. Möller, A. Okem, B. Rama, et al., eds. Cambridge, UK: Cambridge University Press. https://www.ipcc.ch/report/ar6/wg2.

Jones, A.C., and A.J. Lawson. 2021. *Carbon Capture and Sequestration in the United States*. Washington, DC: Congressional Research Service. https://sgp.fas.org/crs/misc/R44902.pdf.

Larson, E., C. Greig, J. Jenkins, E. Mayfield, A. Pascale, C. Zhang, J. Drossman, et al. 2021. *Net-Zero America: Potential Pathways, Infrastructure, and Impacts*. Final Report. Princeton, NJ: Princeton University. https://netzeroamerica.princeton.edu/the-report.

MCC (Mercator Research Institute on Global Commons and Climate Change). 2022. "Remaining Carbon Budget." https://www.mcc-berlin.net/en/research/co2-budget.html.

McCoy, M. 2022. "Green Chemical Maker LanzaTech to Go Public via Merger." *Chemical & Engineering News* March 9. https://cen.acs.org/business/biobased-chemicals/Green-chemical-maker-LanzaTech-to-go-public-via-merger/100/web/2022/03.

Mulligan, J., A. Rudee, K. Lebling, K. Levin, J. Anderson, and B. Christensen. 2020. "Carbonshot: Federal Policy Options for Carbon Removal in the United States." Working Paper. Washington, DC: World Resources Institute. https://files.wri.org/d8/s3fs-public/carbonshot-federal-policy-options-for-carbon-removal-in-the-united-states_1.pdf.

NASEM (National Academies of Sciences, Engineering, and Medicine). 2019. *Negative Emissions Technologies and Reliable Sequestration: A Research Agenda*. Washington, DC: The National Academies Press. https://doi.org/10.17226/25259.

NASEM. 2021. *Accelerating Decarbonization of the U.S. Energy System*. Washington, DC: The National Academies Press. https://doi.org/10.17226/25932.

Olfe-Kräutlein, B., K. Armstrong, M. Mutchek, L. Cremonese, and V. Sick. 2022. "Why Terminology Matters for Successful Rollout of Carbon Dioxide Utilization Technologies." *Frontiers in Climate* 4:830660. https://www.frontiersin.org/article/10.3389/fclim.2022.830660.

Paulos, B. 2021. *Advancing Toward 100 Percent: State Policies, Programs, and Plans for Zero-Carbon Electricity*. Montpelier, VT: Clean Energy States Alliance. https://www.cesa.org/wp-content/uploads/Advancing-Toward-100.pdf.

Sick, V., G. Stokes, and F.C. Mason. 2022. "CO_2 Utilization and Market Size Projection for CO_2-Treated Construction Materials." *Frontiers in Climate* 4(May):878756. https://doi.org/10.3389/fclim.2022.878756.

Tanzer, S.E., and A. Ramírez. 2019. "When Are Negative Emissions Negative Emissions?" *Energy & Environmental Science* 12(4):1210–1218. https://doi.org/10.1039/C8EE03338B.

USCA (United States Climate Alliance). 2019. "Climate Leadership Across the Alliance." 2019 State Factsheets. https://static1.squarespace.com/static/5a4cfbfe18b27d4da21c9361/t/5db99b0347f95045e051d262/1572444936157/USCA_2019+State+Factsheets_20191011_compressed.pdf.

2

Existing Infrastructure for CO_2 Utilization

This chapter describes the existing infrastructure in the United States to capture, transport, and store carbon dioxide (CO_2), and its relationship to current and future CO_2 utilization opportunities. Current sources of CO_2 are also described to assist in evaluating the potential for locating new industries and assessing required infrastructure. Because CO_2 utilization processes and technologies can require inputs of clean electricity, clean hydrogen, water, and natural gas, the status of those industries is also discussed. Net-zero enabling technologies and infrastructure are necessary inputs to directly de-fossilize CO_2 utilization processes themselves and to achieve net-zero emissions of CO_2 utilization systems from a complete life cycle perspective.

2.1 CO_2 SOURCES

Sources of CO_2 for utilization include concentrated point sources of emissions and diffuse resources such as CO_2 in the atmosphere and oceans, particularly collected via direct air capture (DAC) and direct ocean capture (DOC). The following sections describe CO_2 emissions from U.S. sources by location and type of source and provide an overview of the current market for CO_2.

2.1.1 Current U.S. and Global CO_2 Emissions and Stocks

In 2019, 5.259 gigatonnes (Gt) of CO_2 were released into the atmosphere from sources in the United States, equivalent to 1.435 Gt of carbon (EPA 2022). Global emissions of CO_2 amounted to 33.4 Gt CO_2 from the energy system (9.1 Gt carbon), or 40.5 Gt CO_2 (11.0 Gt C) including the energy system and land-use change (Hausfather 2021; IEA 2021). The stock of CO_2 in the atmosphere is about 829 Gt carbon, of which about 240 Gt have been added in the industrial era, when CO_2 began to increase due to human activity, primarily from fossil-fuel combustion (Bruhwiler et al. 2018). By comparison, the global demand for chemicals and derivative materials yielded 0.450 Gt of carbon embodied in the carbon-based products themselves in 2020, 84 percent of which was fossil-derived carbon (57 percent oil, 25 percent natural gas, and 1.7 percent coal), ~10 percent bio-based carbon, and ~5 percent recycled carbon (Kahler et al. 2021). Future CO_2 utilization will aim to replace a portion of the embodied carbon in chemicals and derivative materials with carbon derived from CO_2.

2.1.1.1 Current Point Sources of CO_2

Point sources of CO_2 emissions include electric power plants, industrial and manufacturing facilities, waste facilities, and various other sources of combustion or chemical process emissions. Point sources offer the opportunity to capture emissions before they are released to the atmosphere.

Figure 2-1 depicts CO_2 emissions from major emitting facilities, characterized by source, emission quantity, and location in 2019. The U.S. Environmental Protection Agency (EPA) Facility Level GHG [greenhouse gas] Emissions Inventory (FLIGHT database) facilities are shown in panel (a) (fossil-fuel sources) and panel (b) (industrial facilities), totaling 6,934 facility emission sources with 2,588 million metric tons (MMT) of CO_2 emitted in 2019 (EPA 2017).[1] Fossil facilities include power plants, natural gas and oil systems, and refineries. Industrial emitters include chemical manufacture, industrial gas suppliers, metals, minerals, pulp and paper, waste, and other. The committee also estimated biogenic CO_2 emissions from ethanol plants, panel (c), which are not tracked in the FLIGHT database. Point-source facilities including those shown in Figure 2-1(a)–(c) represent opportunities for

---

[1] CO_2 emissions recorded in the FLIGHT database in 2020 were about 10 percent lower than 2019 emissions, due to the changes in energy use during the COVID-19 pandemic. Because the 2020 data are less representative of the general trend in CO_2 emissions over time, the older 2019 data were used for analysis.

(a)

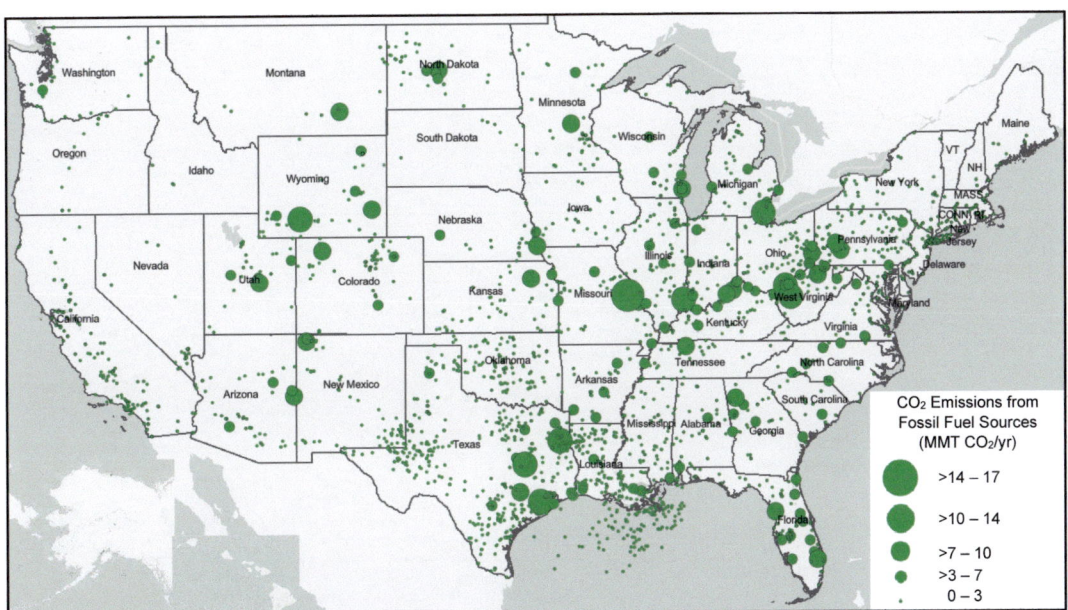

FIGURE 2-1 Map of major facility sources of CO_2 emissions in the United States. Panels (a) and (b) show emissions in 2019 from EPA's FLIGHT database of major emitters, (a) fossil-fuel facilities and (b) industrial emitters. Fossil facilities include power plants, natural gas and oil systems, and refineries. Industrial emitters include chemical manufacture, industrial gas suppliers, metals, minerals, pulp and paper, waste, and other. Panel (c) shows biogenic CO_2 emissions in 2021 from ethanol fermentation plants, estimated from the volume of ethanol produced, using a conversion factor of 0.75 tons of CO_2 emissions from production of 1 ton of ethanol. The circles, squares, and triangles plot the locations of individual sources. The size of the circles, squares, and triangles represents the magnitude of CO_2 emitted by each source.
SOURCE: Data from FLIGHT database for fossil-fuel facility and industrial emissions and from *Ethanol Producer Magazine* at ethanolproducer.com for ethanol plant emissions.

Figure continues on the following page.

(b)

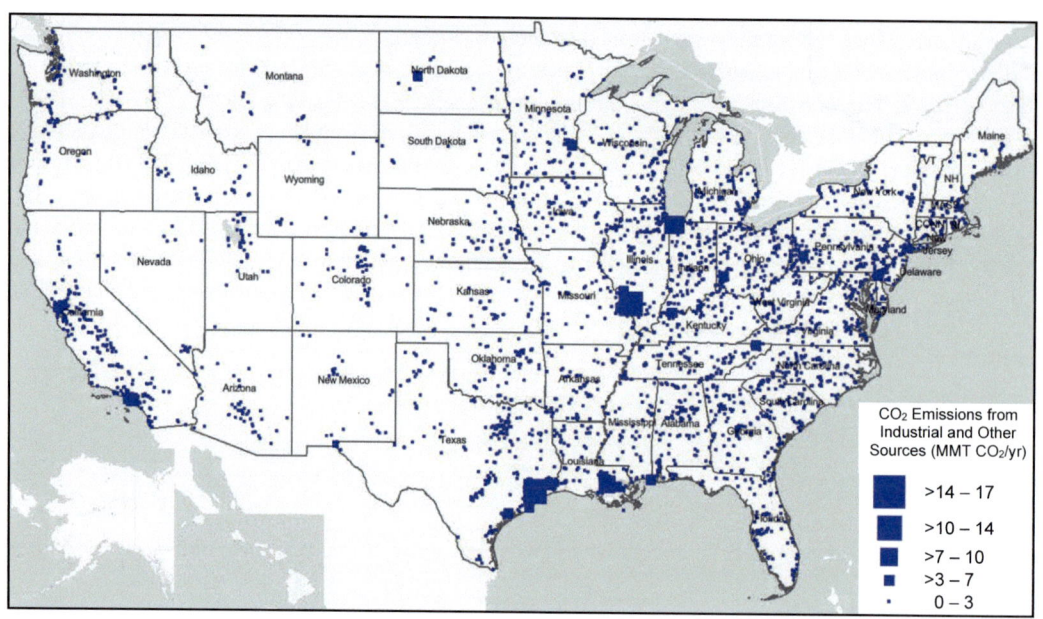

(c)

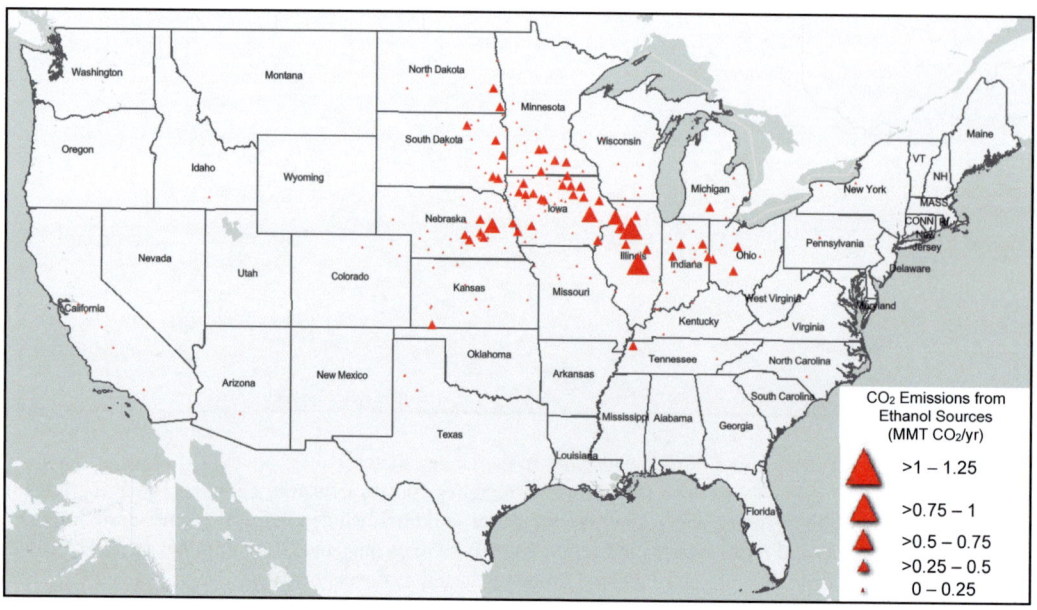

FIGURE 2-1 Continued

emissions capture at the source. Power plants were the largest source of CO_2 emissions among facilities in 2019, followed by facilities from the oil and gas system, refinery, and chemical sectors. The top states for CO_2 emissions were Texas and Louisiana, which have significant emissions from facilities in the top emitting categories.

2.1.1.2 Current Nonpoint Sources of CO_2

Nonpoint sources of CO_2 are those that are difficult to capture at the point of emission, either because they are mobile or too small to warrant local CO_2 capture. CO_2 from these sources likely must be captured from the atmosphere or bodies of water into which the CO_2 dissolves using technologies such as DAC or DOC. With these technologies, the atmosphere and bodies of water could become major reservoirs for carbon that is accessible for use. Major current nonpoint sources of CO_2 include vehicles powered by combustion, distributed agricultural and land-use-based emissions, emissions from combustion-based building heating systems, and natural events such as wildfires. Figure 2-2 plots nonpoint source emissions, not including land-use emissions, compared to point-source emissions.

Transportation is the largest emitting sector in the United States, with 1,818 MMT CO_2 emissions in 2019, representing 35 percent of U.S. CO_2 emissions from combustion of fossil fuels (EPA 2022). Combustion of fuels in homes and businesses for heating and hot water emitted 341 and 251 MMT CO_2, respectively, in 2019. CO_2 emissions related to agricultural and other land uses (not including combustion of fossil fuels) were 7.6 MMT in 2019, primarily from urea and lime use as fertilizers. Land use, land-use change, and forestry currently represent a net CO_2 emissions sink in the United States, with about 800 MMT CO_2 being removed from the atmosphere per year, primarily through forest growth (EPA 2022).

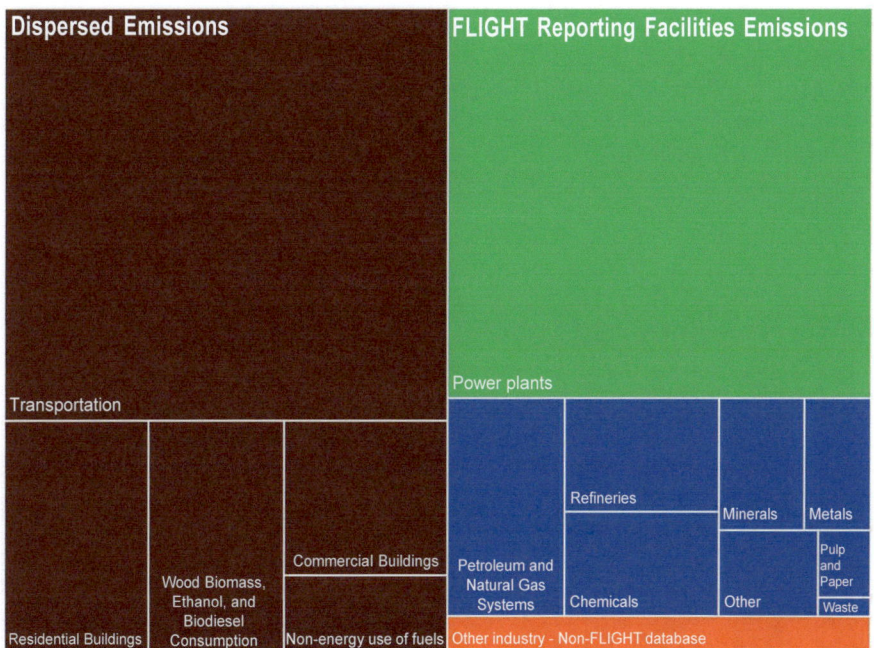

FIGURE 2-2 Total CO_2 emissions in the United States in 2019 not including land-use emissions, showing dispersed emissions (brown) and emissions from major facilities in the FLIGHT database (multicolored). Dispersed emissions from transportation, residential and commercial buildings, biofuel consumption, and non-energy use of fuels could represent a source of CO_2 for utilization if captured via DAC or DOC. Emissions from EPA FLIGHT database reporting facilities for large emitters and other mandated emitters could be captured at the point source. Dispersed emissions are larger than point-source emissions.
SOURCE: Data from EPA, 2022, "Facility Level Information on GreenHouse gases Tool (FLIGHT) Database," and 1999–2019, "Inventory of U.S. Greenhouse Gas Emissions and Sinks."

2.1.1.3 Atmosphere and Water Bodies

DAC and DOC could be used to transform diffuse, atmospheric concentrations of CO_2 into concentrated, point sources of CO_2 for utilization. Currently no commercial facilities for DAC or DOC operate in the United States. However, the number of companies emerging in this space is increasing rapidly, as the newly launched Direct Air Capture Coalition (DAC Coalition 2022) and the 2022 milestone winners in the Carbon Removal XPrize (XPRIZE 2022) show. A recent National Academies report on the research needs for ocean-based CO_2 removal describes a research strategy for various negative emissions aspects of ocean capture (NASEM 2022).

2.1.2 Current Sourcing of CO_2 for Utilization

2.1.2.1 Demand and Supply Balance

CO_2 is a valuable and necessary commodity in today's economy and will become only more integral as new uses emerge. Grand View Research estimated the global market to be over $7.8 billion in 2020 and anticipated growth at a compound annual growth rate of 4.0 percent from 2021 to 2028 (Grand View Research 2022). Global annual use of CO_2 exceeds 230 MMT, primarily in the fertilizer industry for urea manufacturing (130 MMT CO_2), with the second largest user being the oil and gas industry, which consumes 70–80 MMT CO_2 for enhanced oil recovery (EOR; IEA 2019; see Figure 2-3). According to IHS Markit (2021), Mainland China and North America are today's major markets for CO_2, accounting for about 28 percent and 27 percent, respectively, of global demand in 2020, followed by Southwest Asia and the Middle East, accounting for 11 percent and 8 percent, respectively.

The CO_2 market comprises both captive industrial users, who typically produce and consume CO_2 on-site, and merchant users, who receive deliveries of CO_2 in a variety of forms. While the most common captive use of CO_2 today is in the manufacturing of urea, as illustrated in Figure 2-3, various industrial processes, including methanol synthesis, use merchant supply of CO_2. CO_2 is used as an acidifier in beverage and water treatment applications, as a shielding and inerting gas (e.g., in metalworking or food preservation), and as a chilling and cleaning agent while in a solid state. In the merchant market, the food and beverage industry is the largest consumer of liquid CO_2, and this demand is a key external driver in the market.

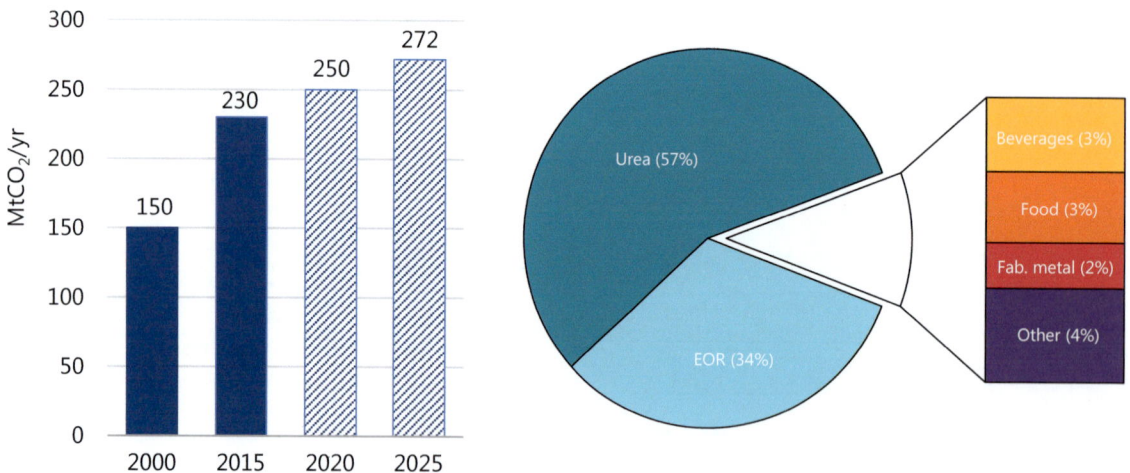

FIGURE 2-3 Current global volume of CO_2 use in Mt CO_2/year. The primary uses of CO_2 globally are in urea production and EOR.
SOURCE: International Energy Agency, 2019, *Putting CO_2 to Use: Creating Value from Emissions*, Paris: IEA, https://www.iea.org/reports/putting-co2-to-use. All rights reserved.

Most merchant CO_2 is recovered in gaseous form from industrial processes with a by-product stream that has a high concentration of CO_2, such as in corn-to-ethanol or hydrogen production. An estimated breakdown of CO_2 sources for merchant supply in the United States includes ethanol (46 percent), natural gas processing (28 percent), NH_3 production (18 percent), steam methane reforming (SMR) plants (6 percent), and nontraditional sources (2 percent). Large amounts of CO_2 also can be extracted from naturally occurring underground sources, such as the McElmo Dome in Colorado or the Jackson Dome in Mississippi (DiPietro et al. 2012). Notably, many of the currently utilized point sources of CO_2 may change in a net-zero emissions future. For example, the demand for bioethanol to blend in gasoline or natural gas for heating and power generation may decrease as vehicles and space heating electrify and the electric grid decarbonizes. Therefore, CO_2 sourced from natural gas processing is likely to decrease as fossil fuels are broadly phased out; however, CO_2 from ethanol production may decrease along with biofuel blend use in transportation, or may remain if ethanol production pivots to other uses of sustainable liquid fuels or carbon products. Depending on the source, the purity of the CO_2 product varies, and merchant CO_2 is often sold under different grades, for example, industrial, food, or specialty grade. Merchant CO_2 is typically liquefied for distribution due to its favorable economics compared to gaseous distribution.

CO_2 market dynamics are variable. Global geopolitics impact the demand for CO_2 for oil and gas production, as observed during the 2020 oil price competition between the Organization of Petroleum Exporting Countries (OPEC) and U.S. shale oil producers, when oil prices plummeted, along with the demand for CO_2 for EOR. The COVID-19 pandemic also led to supply volatility in the United States. Widespread declines in motor vehicle and air travel as people stayed at home reduced demand for fuel, leading to operating reductions or idling of some U.S. ethanol production plants, causing pockets of CO_2 supply shortages (Kelly and Polansek 2020). At the same time, use of CO_2 in the food and beverage sector increased as consumer demand for frozen and packaged foods and carbonated beverages soared. Shipment of food products, as well as vaccine transportation and distribution, boosted the demand for CO_2 in the form of dry ice. Local disruptions in supply-chain markets present an opportunity to find more sources of CO_2 that are more reliable for existing as well as emerging markets.

The major issue in the current CO_2 market is balancing regional supply and demand. CO_2 sources may or may not exist where demand is greatest. As with most industrial gases, production and consumption of CO_2 most often occur within regional networks, due to the high cost of transport relative to the value of the industrial gas product. In the merchant market, CO_2 distribution to end users is often in pressurized cylinders, tankers, and railcars. Overall, current demand for CO_2 is concentrated around population centers, and CO_2 production establishments are usually coupled with or located near downstream customers to reduce transportation costs. Regional production in the United States does not completely parallel population, as illustrated in Figure 2-4. Note that the figure uses production facilities (establishments) as a proxy for production volume, since companies do not release specific production details publicly. Imbalances occur along both coasts, where demand exceeds supply, as well in midwestern and southwestern states, where the reverse condition is observed. Increasing demand for CO_2 can encourage more rapid scale-up and deployment of carbon capture technologies, driving down their costs. The current market for CO_2 is expected to change significantly as uses in the United States develop beyond the existing EOR and industrial processes, to transformations via mineralization, chemical, and biological CO_2 utilization processes.

2.1.2.2 CO_2 Being Captured and Fed into Utilization or Storage

As of April 2022, the United States has 12 commercial-scale carbon capture and storage (CCS) facilities in operation and 58 more in some phase of development or construction (Global CCS Institute 2022). Section 4.1 discusses the infrastructure for CO_2 capture in greater detail. One way to estimate the potential for carbon capture is to determine the number of facilities that could utilize the 45Q tax credit, which, according to a 2020 analysis from the Great Plains Institute, is 1,517 of 6,586 power plants and industrial facilities, accounting for 89 percent of the CO_2 emissions from those sectors (Abramson et al. 2020). As noted above, no commercial U.S.-based facilities for DAC or DOC are operating yet, although a number of research, development, and demonstration (RD&D) projects are ongoing (DOE 2021; NASEM 2022), and the Infrastructure Investment and Jobs Act of 2021 (IIJA) authorized and appropriated $3.5 billion for the creation of four regional DAC hubs, which were defined to

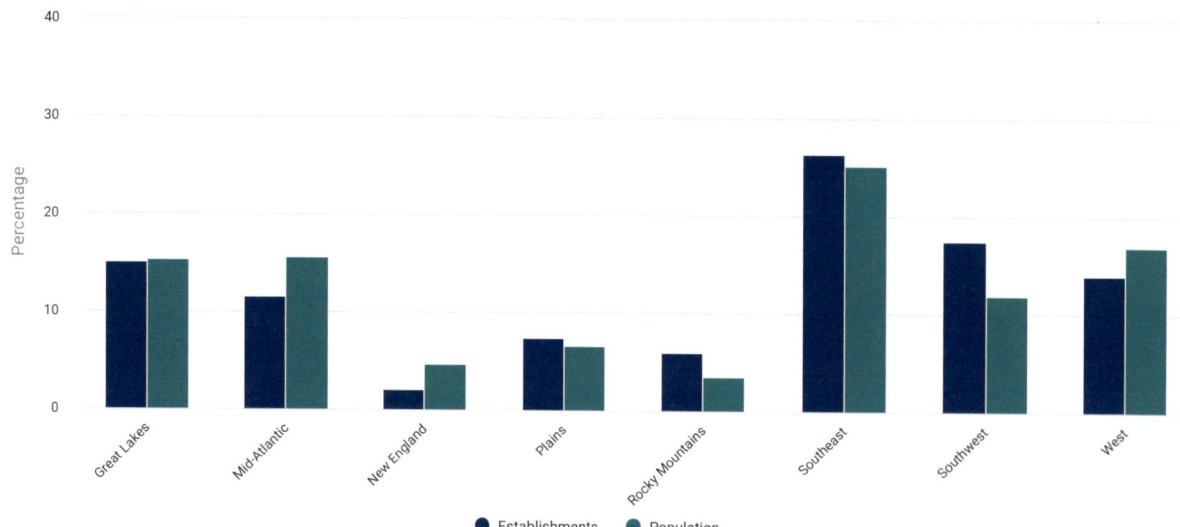

FIGURE 2-4 Distribution of CO_2 production facilities in the United States, including the percentage of CO_2 production facilities (establishments, navy blue) and percent population (teal) of U.S. regions. In general, production parallels population, because CO_2 is relatively expensive to transport. Regional differences in CO_2 production are pronounced based on population, with the Southeast, Southwest, and Great Lakes having the most production. Relative to population, greater production occurs in the Southwest, Rocky Mountains, Southeast, and Plains, as compared to the Great Lakes, West, Mid-Atlantic, and New England.
SOURCE: J. Madigan, 2020, "Carbon Dioxide Production," United States - Specialized Report OD4929, IBISWorld. © IBISWorld.

include capture, potential utilization off-takers, CO_2 transport infrastructure, subsurface resources, and sequestration infrastructure (IIJA 2021 § 40308).

2.2 EXISTING PROCESSES AND FACILITIES UTILIZING CO_2

Established production capability for CO_2-derived products exists, but the number of commercial entities emphasizing new or net-zero CO_2 utilization opportunities has increased only gradually from less than 200 in 2016 (Global CO_2 Initiative 2018), remaining in the low hundreds as of 2022, including CO_2 capture firms (Circular Carbon Network 2022).

2.2.1 Current Mineralization CO_2 Utilization Processes and Facilities

Pilot-scale production of mineralized CO_2 products has demonstrated commercial viability for mineralized waste materials (Hills et al. 2020), and the formation of limestone aggregates has seen early commercial implementation in construction projects (Constantz et al. 2015). CO_2 curing of ready-mix concrete is commercialized (Monkman and MacDonald 2017) and deployed in more than 100 facilities as a bolt-on addition to existing production sites. Precast concrete CO_2 curing and the use of carbonated aggregates in concrete are at advanced stages of deployment (Solidia 2022).

2.2.2 Current Chemical CO_2 Utilization Processes and Facilities

The highest volume chemical transformation of CO_2 today is urea synthesis, as illustrated in Figure 2-3, above. Figure 2-5 shows other major products chemically synthesized from CO_2 at industrial scale, which include organic

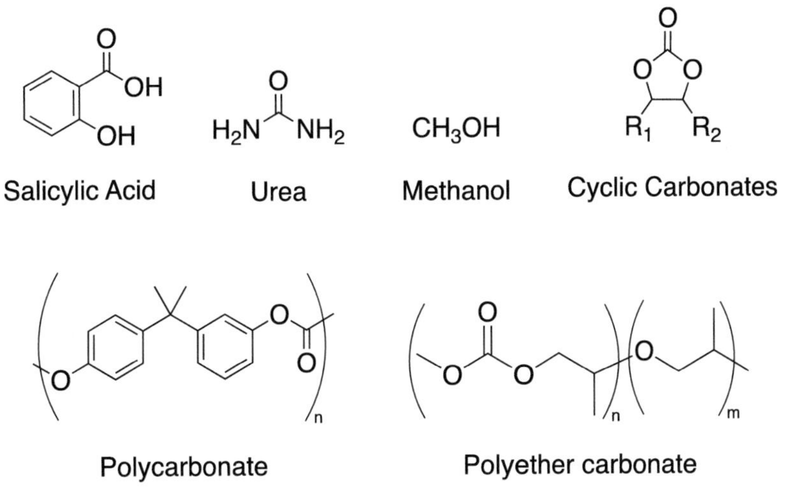

Chemical	Scale per year (tons)[a]
Salicylic acid	90,000
Urea	175,000,000
Methanol	2,000,000
Cyclic carbonates	40,000
Polycarbonates	600,000
Polyether carbonates	10,000

[a] Amount produced using a process involving CO_2.

FIGURE 2-5 Chemicals synthesized on a large scale from CO_2.
SOURCE: Adapted with permission from M.D. Burkart, N. Hazari, C.L. Tway, and E.L. Zeitler, 2019, "Opportunities and Challenges for Catalysis in Carbon Dioxide Utilization," *ACS Catalysis* 9(9):7937–7956, https://doi.org/10.1021/acscatal.9b02113. Copyright 2019, American Chemical Society.

carbonates and their polymers, methanol, and salicylic acid. The scale of urea production dwarfs that of the other chemicals by at least two orders of magnitude. These chemicals are produced using CO_2 because it is chemically advantageous to do so, in part because they are relatively highly oxygenated.

In addition to industrial-scale transformation of CO_2, pilot-scale utilization is emerging for additional processes and products. These processes and products include carbon monoxide, methanol, methane, other hydrocarbon fuels, oxalic acid, ethylene carbonate, ethylene glycol, and polyurethane (Bright 2020; Burkart et al. 2019). These pilot projects are implementing known CO_2 transformation technologies with sustainable CO_2, or are developing new transformations of CO_2, each of which are important steps in further developing commercial CO_2 utilization. Methanol from CO_2 is the most advanced of the emerging chemical CO_2 utilization technologies, with major pilot projects and commercial-scale plants in the works in both Europe and Asia (Carbon Recycling International n.d.; Greenwood 2021; Hobson 2018; Mitsubishi Gas Chemical Company 2021; Pires da Mata Costa et al. 2021). Likewise, pilot plants for carbon monoxide production using solid oxide electrolyzer technology have led to commercially available electrolyzers (Küngas et al. 2017). There are multiple pilot plants of various sizes in Europe for producing methane and Fischer-Tropsch fuels, which have market opportunities in storing energy and powering aviation (Ineratec 2017). As one example, Haldor-Topsoe is partnering with Arcadia eFuels in Denmark to implement their gas-to-liquid "e-fuels" technology in a commercial-scale plant, which is scheduled to start operations by the end of 2024 (Arcadia eFuels 2022; Haldor-Topsoe 2022). Several companies have also commercialized technologies for the production of polycarbonates from CO_2; as one example, Asahi Kasei's CO_2 to polycarbonate process is being used worldwide to produce around 900,000 tons of polycarbonate per year, equivalent to 16 percent of total polycarbonate production capacity (Asahi Kasei Corporation 2021). Pilot plants to utilize CO_2 in producing oxalic acid, ethylene glycol, and polyurethane, among other chemicals and plastics, are all below 50 tonnes of product per year, and thus are still pre-commercial (Scott 2019; Yang et al. 2019).

2.2.3 Current Biological CO_2 Utilization Processes and Facilities

CO_2 utilization by microbes (naturally occurring or genetically modified) ranges from the production of chemicals as simple as ethanol, for which large-scale facilities exist, all the way to complex molecules that include proteins. In particular, green algae are being used to produce biodiesel, renewable synthetic diesel and gasoline, protein, pentanoic acids, and astaxanthin. Cyanobacteria yield ethanol, *iso*- and *n*-butanol, butanediol, propanediol, hydroxybutanoic and propanoic acids, fatty acids, ethylene, 2-methyl-1,3-butadiene, heptadecane, limonene, bisabolene, squalene, and farnesene (Burkart et al. 2019). Some products are made as mixtures, such as biodiesel. Several biological CO_2 utilization processes are in the pilot stage. Most biological processes for making carbon-based chemicals and materials use CO_2 indirectly, via plants using CO_2 to make biomass, conversion of biomass to sugar, and then sugar to the final product. This study examines only direct uses of CO_2 by microbes for CO_2 utilization.

2.3 EXISTING CO_2 TRANSPORT AND STORAGE INFRASTRUCTURE

2.3.1 CO_2 Pipeline Systems

As of 2020, there were 5,150 miles of pipelines for supercritical fluid CO_2 transport in the United States, controlled by 32 different pipeline operators, with the largest pipeline networks in the Gulf Coast, Permian Basin, and Rocky Mountain areas (CEQ 2021; NPC 2021b; PHMSA 2022c). An additional 60 miles of CO_2 pipelines transport CO_2 in the gaseous state (CEQ 2021). The existing CO_2 pipeline network is shown by the yellow lines in Figure 2-6. Pipelines for transporting CO_2 as a supercritical fluid have the capacity to move 890 to 103,000 tonnes of CO_2 per day depending on their length and diameter and must be maintained above 1,080 psi (NPC 2021b).

Approximately 90 percent of the CO_2 pipeline infrastructure in the United States is used for EOR and primarily connects geologic CO_2 sources with depleted oil reservoirs (Edwards and Celia 2018; NPC 2021b). As of the end of 2020, U.S. EOR operations used 1.6 billion cubic feet of CO_2 per day (Bcf/d), 1.3 Bcf/d from geologic sources and 0.3 Bcf/d from industrial sources (ARI 2021), which is equivalent to about 67,000 tonnes/day from geologic sources and 16,000 tonnes/day from industrial sources. EOR is not within the scope of CO_2 utilization as defined in this report (see Box 1-2). Nonetheless, the existing infrastructure and over 50 years of experience transporting CO_2 for EOR can be leveraged when developing future CO_2 utilization infrastructure.

Pipeline transport has the benefit of achieving economies of scale for large volumes of CO_2, but its development has high capital costs (NPC 2021b). Using the National Energy Technology Laboratory's CO_2 transport cost model, the Great Plains Institute estimated costs for capital investment as well as operation and maintenance of CO_2 pipelines, finding that large, shared trunk lines transporting greater than 12 million tonnes CO_2 per year could cost less than \$10/tonne CO_2, while small feeder lines transporting less than 4 million tonnes of CO_2 per year could cost in excess of \$20/tonne CO_2 (Abramson et al. 2020).[2] Actual costs for CO_2 pipeline construction have varied widely due to geographic and right-of-way considerations for different projects (e.g., costs of acquisition, permitting, and construction). For example, pipeline costs for six projects completed since 2009 ranged from \$68,635 to \$199,176 per diameter inch mile (NPC 2021b). In general, an

---

[2] See the appendix of Abramson et al. (2020) for details about the methodology used to make these cost estimates.

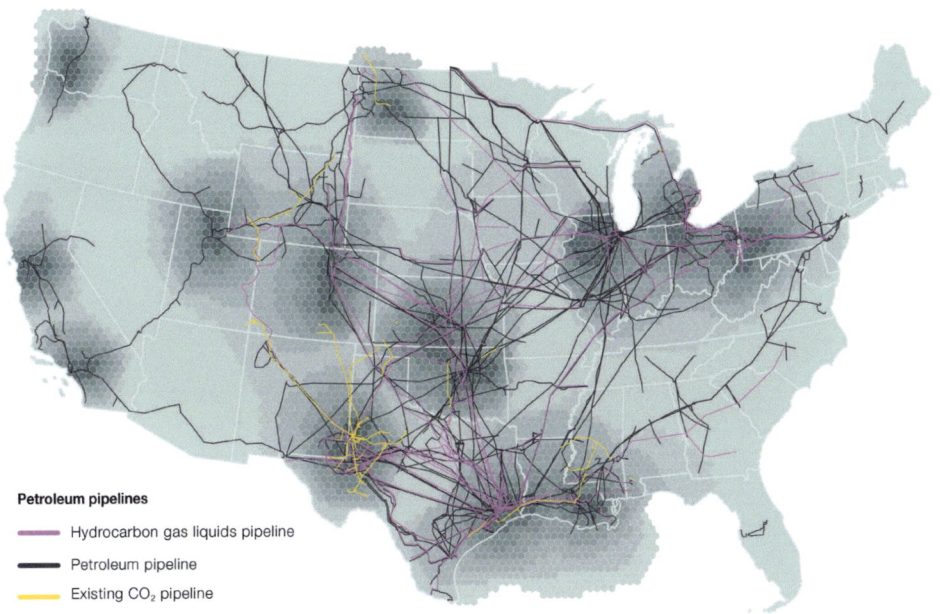

FIGURE 2-6 Map of existing hazardous liquid pipeline networks for CO_2 (yellow), hydrocarbon gas liquids (pink), and petroleum (gray) transport in the contiguous United States.
NOTES: Hydrocarbon gas liquids (HGLs) are defined as hydrocarbons that exist in the gas phase at atmospheric pressure and in the liquid phase at higher pressures. HGLs include alkanes such as ethane, propane, butane, and pentane, and alkenes such as ethylene, propylene, and butylene (EIA 2021).
SOURCE: Adapted from E. Abramson, E. Thomley, and D. McFarlane, 2022, *An Atlas of Carbon and Hydrogen Hubs for United States Decarbonization*, Minneapolis, MN: Great Plains Institute, https://scripts.betterenergy.org/CarbonCaptureReady/GPI_Carbon_and_Hydrogen_Hubs_Atlas.pdf.

inverse relationship between cost per tonne CO_2 and volume of CO_2 transported is expected and argues for coordinated infrastructure development (Abramson et al. 2020). Oversizing of pipeline infrastructure is more cost-effective in the long term because it can allow for future expansion without the environmental impacts of an additional pipeline project. Section 6.1 provides more information about funding and investment opportunities for shared pipeline development.

2.3.2 Other Transportation Methods

Alternative modes of CO_2 transport include truck, rail, ship, and barge. Figure 2-7 shows the rail lines used in the contiguous United States for fuel transport, and Figure 2-8 shows U.S. interstates and navigable waterways. Truck and rail transport are most viable for shorter distances (less than 200 miles for truck and less than 1,000 miles for rail) and for smaller CO_2 volumes (up to several hundred tonnes) but typically cost about 3–10 times more per tonne CO_2 than pipeline transport (NPC 2021b). Currently, the beverage industry is the primary user of truck CO_2 transport, carrying 18 tonnes of CO_2 per truck (NPC 2021b). Ship and barge transport can be used for crossing large bodies of water or navigating intercoastal waterways, respectively; neither option is used widely as of the writing of this report, though ship transport is gaining interest in Europe, notably as part of the Northern Lights project in Norway (Northern Lights 2021; NPC 2021b). For CCS applications, ship transport could provide some flexibility for use at either onshore or offshore storage locations (ZEP 2011). Ship transport can also be cost-effective, but siting pressurized CO_2 shipping and receiving terminals requires consideration of public safety concerns (NPC 2021b).

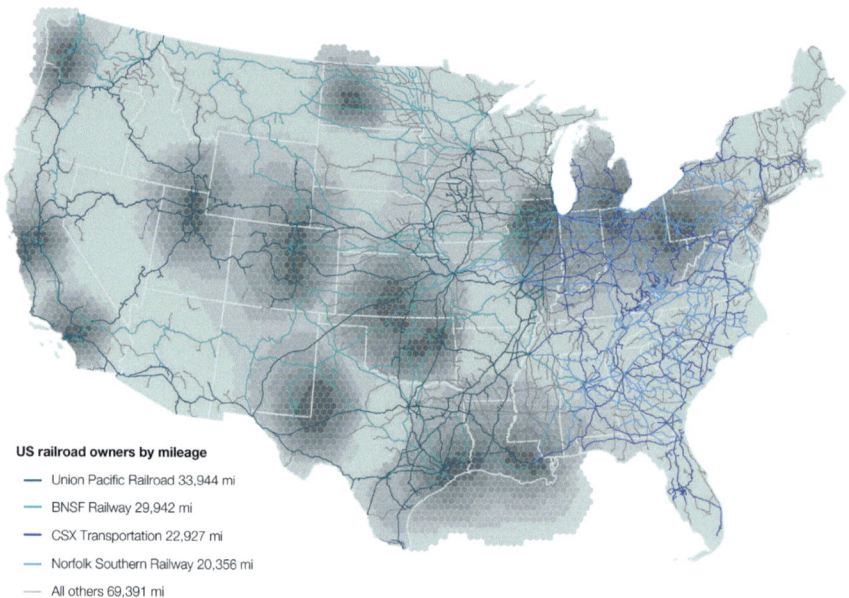

FIGURE 2-7 Map of railroads used for fuel transport in the contiguous United States.
SOURCE: E. Abramson, E. Thomley, and D. McFarlane, 2022, *An Atlas of Carbon and Hydrogen Hubs for United States Decarbonization,* Minneapolis, MN: Great Plains Institute, https://scripts.betterenergy.org/CarbonCaptureReady/GPI_Carbon_and_Hydrogen_Hubs_Atlas.pdf.

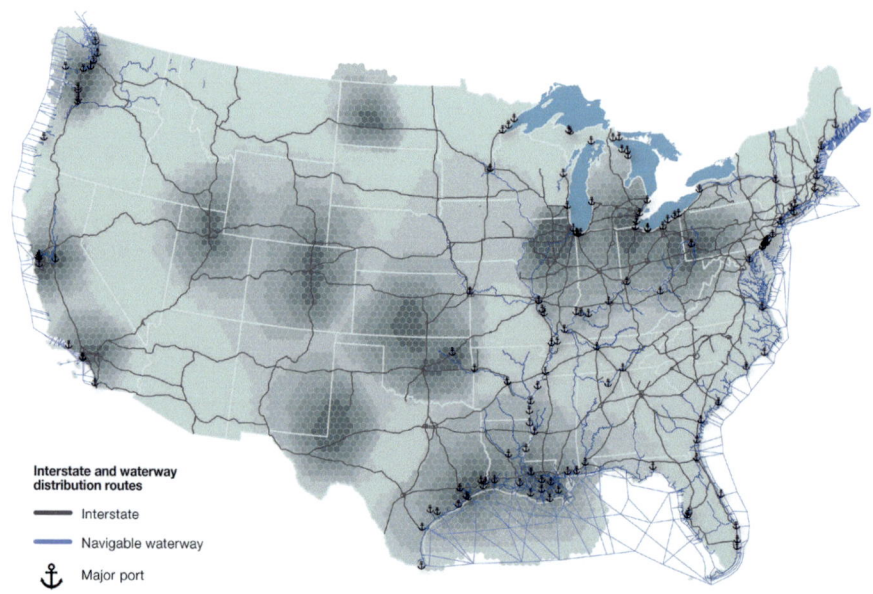

FIGURE 2-8 Map of interstates (gray) and navigable waterways (blue) available for CO_2, hydrogen, or other fuel transport in the contiguous United States, with major ports indicated by an anchor symbol.
SOURCE: E. Abramson, E. Thomley, and D. McFarlane, 2022, *An Atlas of Carbon and Hydrogen Hubs for United States Decarbonization,* Minneapolis, MN: Great Plains Institute, https://scripts.betterenergy.org/CarbonCaptureReady/GPI_Carbon_and_Hydrogen_Hubs_Atlas.pdf.

2.3.3 Co-located CO_2 Capture and Use

Currently, no commercial-scale facilities in the United States co-locate capture and use of CO_2 for purposes other than EOR or urea production. Some government–industry collaborations test and research real-world operating conditions for co-located CO_2 capture and utilization (CCU), for example, at Wyoming's Integrated Test Center and the National Carbon Capture Center (National Carbon Capture Center 2022; Wyoming ITC n.d.). The XPRIZE Foundation's NRG COSIA Carbon XPRIZE supported testing of breakthrough CO_2 utilization technologies, providing the competition's finalist teams with access to power plant flue gas and testing bays equipped with other necessary inputs such as electricity and water (XPRIZE n.d.). The U.S. Department of Energy (DOE) is providing funding for two front-end engineering design (FEED) studies to integrate DAC with CO_2 conversion technologies, one to generate concrete and the other to electrochemically produce formic acid (DOE-FECM 2022).

Also under development are systems for "capture and conversion," or "reactive capture," which integrate CO_2 separation from waste gas and conversion to product in a single step. Whereas traditional CO_2 "capture and release" is limited by the need for weak capture binding energy that results in lower capture efficiency, "capture and conversion" enables stronger binding free energies for the capture step, resulting in more efficient systems that require less sorbent material and energy (Omodolor et al. 2020; Shao et al. 2022). Desorption is accomplished by first reacting the captured CO_2, for example, with hydrogen or electrochemically, producing a less strongly sorbed product that can be released at low energy cost, while also forming a higher-value-added product using the same sorbent and system ("process intensification"). Capture and conversion technology is much less mature (technology readiness level <5) than traditional carbon capture but is an active area of research to reduce costs for CCU (Deutsch et al. 2021).

Co-location of CO_2 capture and conversion facilities provides the potential to reduce transportation costs and energy footprints, and can synergistically align regulatory requirements. These opportunities must be identified on a case-by-case basis, aligning CO_2 source (fossil versus biogenic or atmospheric) with duration of product end use or end-of-life sequestration, to ensure sustainable outcomes. Project developers will need to consider the future availability of CO_2 point sources, particularly fossil sources, before making investments in capital-intensive conversion processes and infrastructure. These challenges in some cases may require flexible and adaptive enabling infrastructure to match distributed electricity, hydrogen production, and other inputs with distributed CO_2 capture and utilization processes, and end-use applications.

2.3.4 CO_2 Storage Infrastructure

To achieve a net-zero emissions future, the endpoint of most captured CO_2 probably will need to be long-term geologic storage.[3] As noted above, much of the existing CO_2 pipeline infrastructure has been developed to inject CO_2 underground for EOR. Only 1 of the 12 commercial CCS facilities in the United States is currently injecting CO_2 *solely* for the purpose of storage (and not for EOR): the Archer Daniels Midland plant in Illinois, which captures and stores about 3,000 tonnes of CO_2 per day from a bioethanol plant (Jones and Lawson 2021; NPC 2021a). The other 11 facilities inject CO_2 only for EOR. For secure underground storage of CO_2, the geology must include a storage reservoir with the porosity to hold CO_2 and the permeability to accommodate injection, situated underneath an impermeable rock formation to minimize CO_2 leakage (LEP 2021). Geologic formations considered for storage include depleted oil and gas reservoirs, deep saline reservoirs, and unmineable coal seams. Estimates of storage capacity in the United States range from 2.6 to 22 trillion tonnes CO_2, pending future evaluation of the suitability of sites for proximity to CO_2 sources, porosity, permeability, and potential for leakage (Jones and Lawson 2021). Figure 2-9 shows potential CO_2 storage sites in the United States and their proximity to existing CO_2 pipelines, indicating those sites that are expected to have very low (green) and low (yellow) storage costs,

---

[3] Best practices have been established for monitoring and verification of CO_2 storage to ensure safety and permanence. These involve considerations of risks that could result in CO_2 leakage over the long term (e.g., hundreds of years) (NETL 2017a). As one example, use of depleted hydrocarbon reservoirs for CO_2 storage can mitigate seismicity risks related to CO_2 injection (Dvory and Zoback 2021). Financial risks and liabilities that extend beyond a CCS project's operational lifetime also need to be considered and addressed (Donlan and Trabucchi 2011), but doing so is beyond the scope of this study.

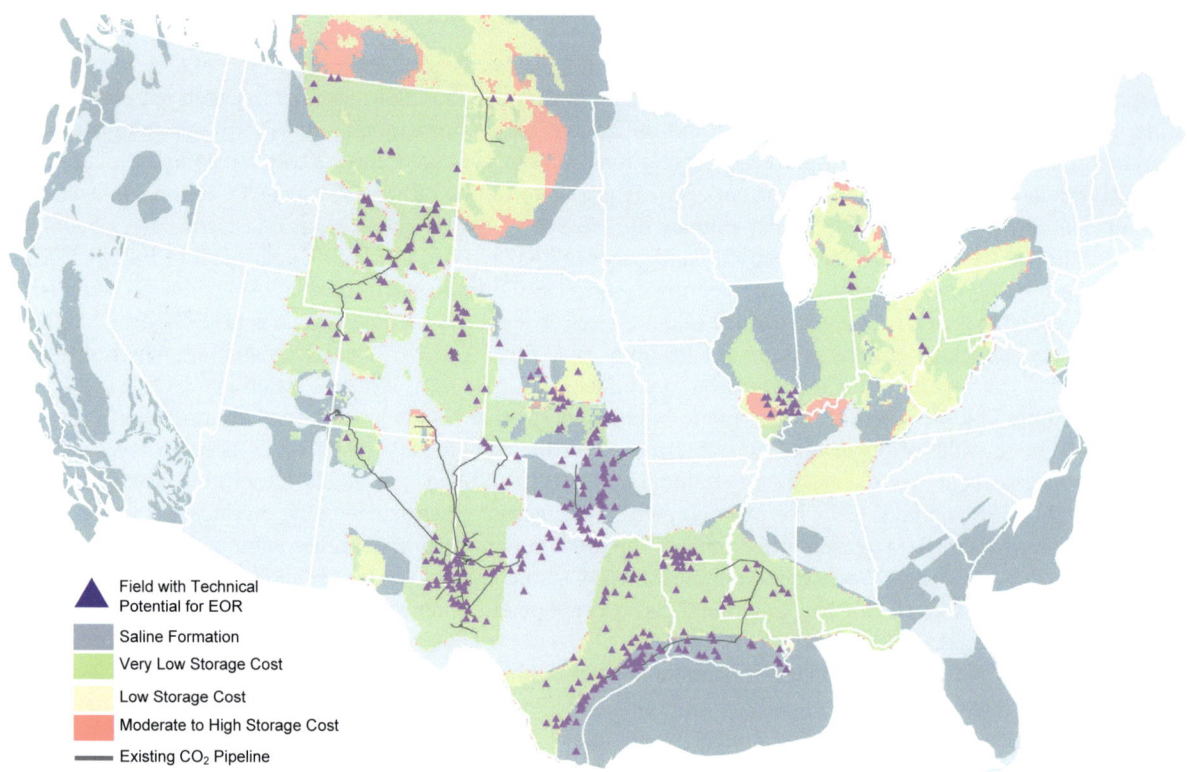

FIGURE 2-9 Map of potential CO_2 storage sites in the contiguous United States, showing existing CO_2 pipelines (gray lines) and possible EOR locations (purple triangles) as well as areas projected to have very low (green), low (yellow), and moderate-to-high (red) storage costs.
SOURCE: Adapted from E. Abramson, D. McFarlane, and J. Brown, 2020, *Transport Infrastructure for Carbon Capture and Storage Whitepaper on Regional Infrastructure for Midcentury Decarbonization*, Minneapolis, MN: Great Plains Institute, https://www.betterenergy.org/wp-content/uploads/2020/06/GPI_RegionalCO2Whitepaper.pdf.

respectively. DOE's cost estimates for CO_2 storage in the United States range from \$7 to \$13/t CO_2 (2013\$), depending on the depth of the geologic formation and the number of injection wells needed, the volume and purity of the CO_2 for storage, the current use of the land, and the capability for surface and subsurface monitoring (NPC 2021a). Detailed discussion of CO_2 storage costs is outside the scope of this report, but more information on CCS costing methodologies can be found in Rubin et al. (2013), and estimates for specific projects can be made using the National Energy Technology Laboratory's CO_2 Saline Storage Cost Model (NETL 2017b).

2.4 STATUS OF ENABLING INFRASTRUCTURE FOR CO_2 UTILIZATION

Enabling infrastructure for CO_2 utilization encompasses the sourcing and transport of other feedstocks and components needed to manufacture CO_2-derived products. An important aspect of project development is considering the energy and fuel consumption requirements for capturing CO_2. As will be further discussed in Chapter 3, clean energy and hydrogen are required in large amounts to upgrade CO_2 into hydrocarbon products. Thus, expansion of CO_2 utilization for products encompassing large sectors of the U.S. economy, including fuels and chemicals, would require extensive expansion of infrastructure for clean electricity and hydrogen. Water may be required (e.g., for on-site generation of hydrogen or process cooling), and wastewater may be generated that requires treatment. This section overviews the status of enabling infrastructure for CO_2 utilization, specifically clean electricity, hydrogen, water, and natural gas.

2.4.1 Clean Electricity

The electrical grid in the United States is undergoing a transformation from majority fossil-based to predominantly low-carbon sources, including renewable wind and solar, hydro, geothermal, nuclear, and natural gas or biomass with CCS. In 2021, 61 percent (2,504 billion kilowatt hours [kWh]) of U.S. electricity generation came from fossil sources, 19 percent (778 billion kWh) came from nuclear, and 20 percent (826 billion kWh) came from renewables (wind, hydropower, solar, geothermal, and biomass) (EIA 2022). Figure 2-10 shows the locations of non-fossil electricity generating facilities in the United States. Clean electricity will become the most important vector for decarbonizing energy-consuming sectors, including transportation, building heating and hot water, and industry. Electricity demand will increase significantly to achieve net-zero goals. For example, converting from petroleum-fueled vehicles to highly efficient battery electric vehicles will shift energy consumption from petroleum to electricity, placing increasing demand on grid utilization. Changes to manufacturing that replace thermally driven processes with electrically driven ones will further increase electricity demand. Projections of electricity demand for a net-zero economy in the United States range from 6,000 to 10,500 terawatt hours (TWh) in 2050, depending on the assumptions used in modeling; regardless, all represent significant increases from the current electricity demand, approximately 4,000 TWh in 2019 (Holmes et al. 2021).

Current U.S. roadmaps for electricity system expansion include electrification of transportation and in some cases residential heating, but generally lack significant consideration of electrification needs for industrial process decarbonization (see, e.g., Larson et al. 2021; Murphy et al. 2021). Many industrial processes currently proposed for CO_2 utilization are capital intensive, given the complexity of the facilities, the multiple steps involved, and/or the high operating temperatures and pressures. In these cases, the most economically efficient and safest operational profile would be 24/7. Grid interconnection and energy storage therefore can become important enablers to maintain constant operation with net-zero carbon electricity inputs, given the intermittency of appropriately sized renewable energy generation. Geothermal energy and hydroelectric power can provide dispatchable renewable electricity but are limited in location and scope. Long-term energy storage is expected to be particularly important

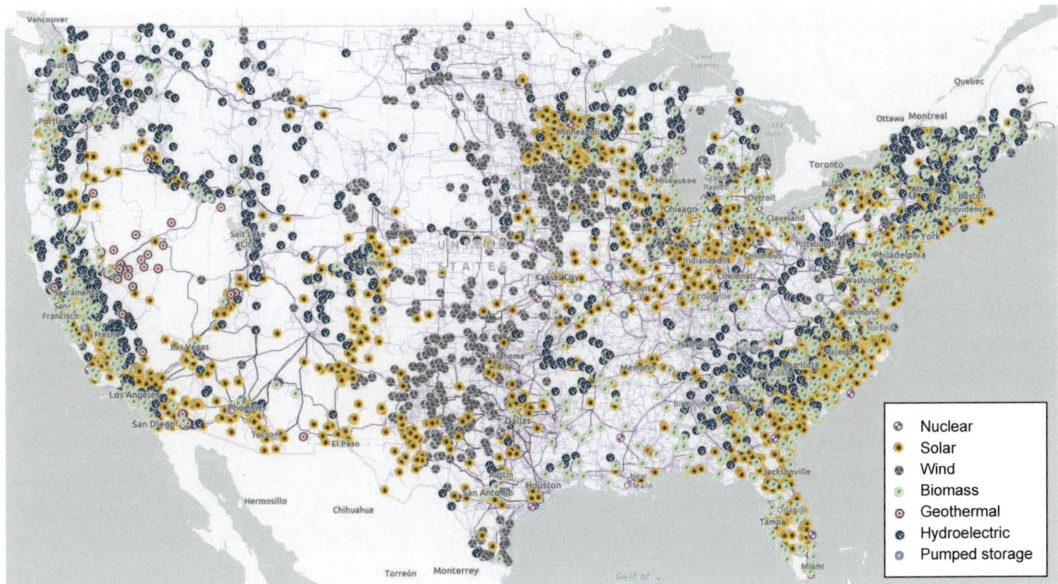

FIGURE 2-10 Location of non-fossil energy generating sources and transmission lines in the United States as of December 2021.
SOURCE: Adapted from U.S. Energy Information Administration, 2021, "Map of the Electricity Infrastructure in the U.S.," https://atlas.eia.gov/apps/electricity/explore.

in future energy systems with significantly increased power generation from wind and solar (Dowling et al. 2020; Shaner et al. 2018). Without energy storage, simpler low-capital-expenditure processes operating with few process steps, at lower pressures, and requiring less heat integration may be considered if electricity inputs are intermittent. Potential examples of such processes include some mineralization or biological CO_2 conversion approaches that are envisioned as having lower capital costs; however, there are limited operating examples, so this possibility has yet to be proven. Historically in its FutureGen and Vision21 programs, and today in its Integrated Energy Systems Program, DOE has studied operational strategies for power plants that use power at times of low electricity demand to make storable fuels. Learning from these programs can be leveraged for potential intermittent use of electricity by CO_2 utilization projects.

2.4.2 Hydrogen

Hydrogen will be an important input to many CO_2 utilization processes. The United States produces approximately 11 million tonnes of hydrogen annually. Current primary uses of hydrogen include petroleum refining to fuels and chemicals, food processing (e.g., hydrogenation of vegetable oils), and production of fertilizer. Studies estimate that for hydrogen to serve as a clean energy vector and contribute to achieving net-zero goals, a five-fold or more increase in its use may be required (FCHEA 2020; Ruth et al. 2020; World Energy Council 2021). Consequently, a substantial expansion in production, storage, and distribution facilities also would be needed. The majority (~77 percent) of hydrogen produced annually in the United States is generated via SMR, with 23 percent produced as a by-product of petroleum refining and other industrial processes and less than 1 percent from water electrolysis (EFI 2021). Use of hydrogen in a future net-zero emissions system will require decarbonization of production. Generation of clean hydrogen from natural gas with CCS has been demonstrated at commercial scale (Global CCS Institute 2021), and advanced technologies (e.g., autothermal reforming and partial oxidation) are known commercially to integrate better with carbon capture than existing technologies. Another method for producing clean hydrogen is water electrolysis using low-carbon electricity; as of May 2022, there were 18.5 megawatts (MW) installed electrolysis capacity and 602.6 MW planned capacity located across the United States (DOE 2022; see Figure 2-11).[4] Methane pyrolysis to generate hydrogen and solid carbon, with no CO_2 emissions, is the subject of ongoing research and development with one facility having achieved commercial scale as of June 2022 (PARC 2022).

Much hydrogen is used on the site of production. When required, hydrogen transport from the site of production to the site of use occurs mostly by pipeline, or alternatively via tube trailer and tanker truck or, for larger volumes, by rail or ship (DOE 2020). The United States currently has about 1,600 miles of hydrogen pipelines supported by a private pipeline network primarily located in the Gulf Coast (DOE-EERE n.d.), as shown in Figure 2-12. This existing network could provide some connectivity and rights-of-way for future expansion. Gaseous hydrogen delivery by tube trailer is currently a common method to match supply and demand in relatively small quantities, about 1–2 tonnes per tube trailer. Tanker trucks can accommodate about five times more hydrogen than tube trailers, given that they transport hydrogen in liquid form, which maintains purity better than gaseous hydrogen transport (DOE 2020). However, liquefying hydrogen for transport in tanker trucks has significant energy requirements: state-of-the-art liquefaction systems add about 12 kWh/kg to the energy requirements for hydrogen production and storage, which represents about one-third of the energy required to produce hydrogen (33 kWh/kg) at the limit of 100 percent efficiency (IDEALHy n.d.). Industry and DOE research targets aim ultimately to reduce the specific power consumption to 6 kWh/kg (DOE-HFTO n.d.). In some cases, hydrogen requires storage before or after transport; its storage as a compressed gas (350 to 700 bar, or about 5,080 to 10,150 psi) and as a cryogenic liquid are well known but require additional development and deployment to decrease costs and increase capacity (DOE 2020). As further discussed in Chapter 4, for some CO_2 utilization applications, on-demand, on-site hydrogen production may be preferable to avoid the costs and challenges of transport and storage.

---

[4] In the upper limit of production, assuming a 100 percent capacity factor and 100 percent efficient electrolyzer, the installed electrolysis capacity would generate about 4,860 tonnes H_2 per year, and the installed plus planned capacity would generate around 163,000 tonnes H_2 per year.

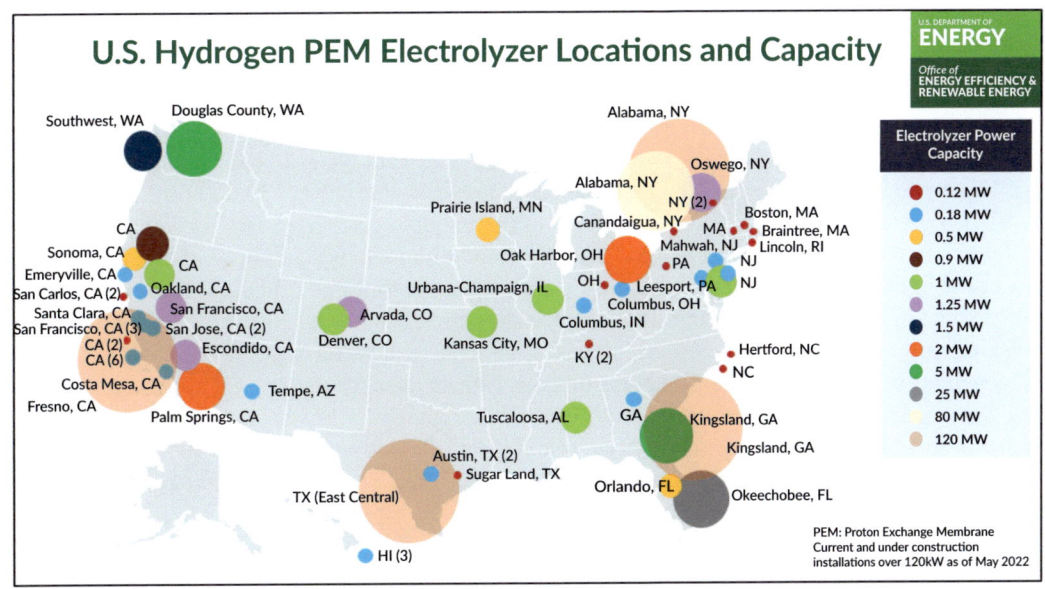

FIGURE 2-11 Current and planned electrolyzer location and capacity (megawatts) in the United States as of May 2022.
SOURCE: U.S. Department of Energy, 2022, "PEM Electrolyzer Capacity Installations in the United States," DOE Hydrogen Program Record 22001, https://www.hydrogen.energy.gov/pdfs/22001-electrolyzers-installed-in-united-states.pdf.

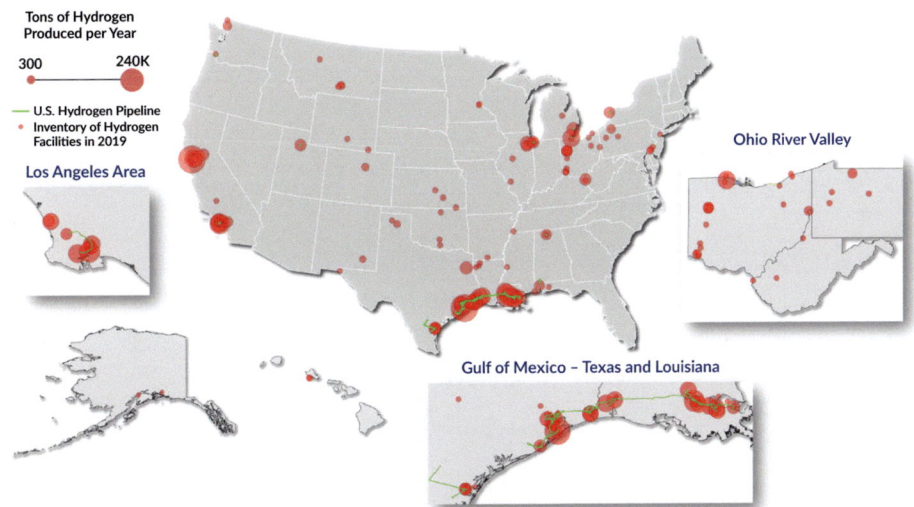

FIGURE 2-12 Hydrogen production facilities (red circles) and pipelines (green lines) in the United States as of 2021. The size of the red circles represents the amount of hydrogen produced at the facility per year in tons, ranging from 300 to 240,000.
NOTES: Figures 2-11 and 2-12 use different units to depict the amount of hydrogen production, megawatts and tons per annum, respectively. Converting from megawatts to tons of hydrogen requires knowing the capacity factor and electrolyzer efficiency; in the upper limit, assuming a 100 percent capacity factor and 100 percent efficient electrolyzer, a 1-MW electrolyzer would produce 290 tons (260 tonnes) of hydrogen in 1 year.
SOURCE: Energy Futures Initiative, 2021, *The Future of Clean Hydrogen in the United States: Views from Industry, Market Innovators, and Investors*, Part of the EFI Report Series *From Kilograms to Gigatons: Pathways for Hydrogen Market Formation in the United States*, Washington, DC, https://energyfuturesinitiative.org/wp-content/uploads/sites/2/2022/03/The-Future-of-Clean-Hydrogen-in-the-U.S._Report-1.pdf.

2.4.3 Water

Water is a required input for several aspects of CCU, including for CO_2 capture, hydrogen generation, and production of the utilization product, especially for biological CO_2 utilization processes that need water for algae and other microbe cultivation. Water use in the United States is documented by the U.S. Geological Survey; their most recent assessment in 2015 estimated self-supplied industrial water use at 14,800 million gallons per day, of which 95 percent is freshwater, and thermoelectric power water use at 133,000 million gallons per day, of which 87 percent is freshwater (Dieter et al. 2018). The energy sector accounts for approximately 10 percent of U.S. water consumption and 40 percent of U.S. water withdrawals (Grubert and Sanders 2018). Figure 2-13 shows total water withdrawals by state, with a darker blue coloration indicating higher water withdrawals. Notably, some of the highest-withdrawal states, for example, Texas and California, are also those with the greatest risk of water stress, a metric comparing water demand with water supply (see Figure 2-14). In these locations, desalination may be required to obtain sufficient freshwater for the desired application. Desalination costs need to be considered in the capital expenditures but are generally lower than the impact of electrolyzer capital expenditures or the cost of electricity; as an example, the additional energy requirements to perform desalination by reverse osmosis are estimated to result in added costs of $0.01–0.02 per kg hydrogen produced by electrolysis (Beswick et al. 2021).

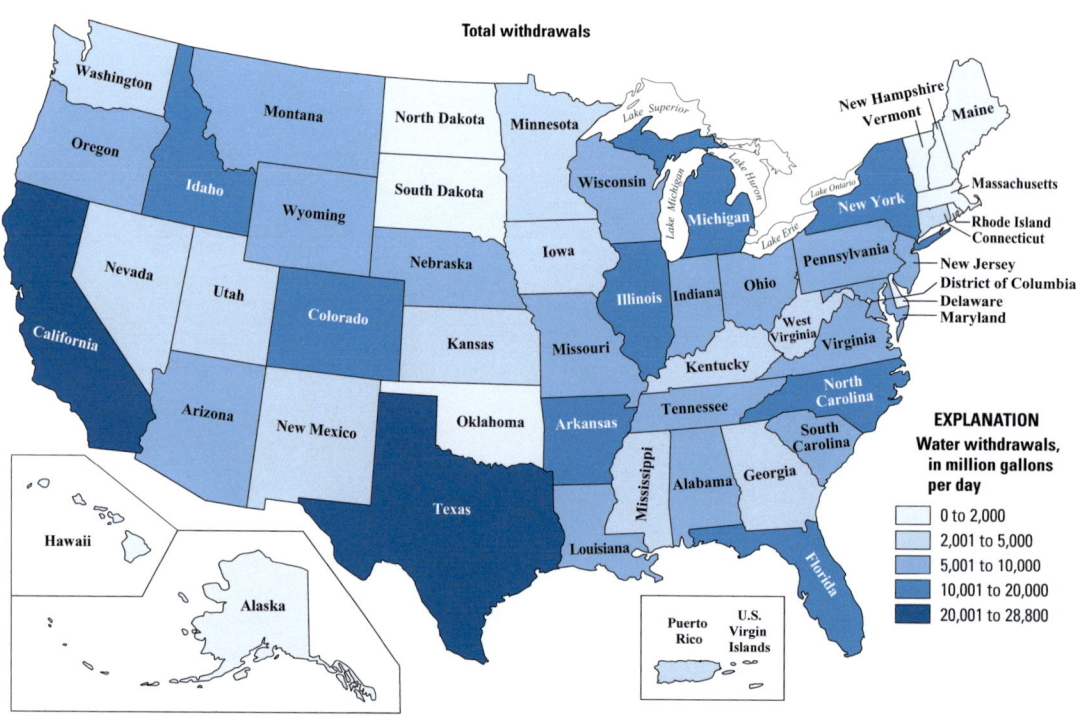

FIGURE 2-13 Total water withdrawals in the United States in 2015, where darker blue indicates higher water withdrawals. SOURCE: C.A. Dieter, M.A. Maupin, R.R. Caldwell, et al., 2018, *Estimated Use of Water in the United States in 2015*, Water Availability and Use Science Program, Circular 1441 (supersedes USGS Open-File Report 2017-1131), Reston, VA: U.S. Geological Survey, https://pubs.usgs.gov/circ/1441/circ1441.pdf.

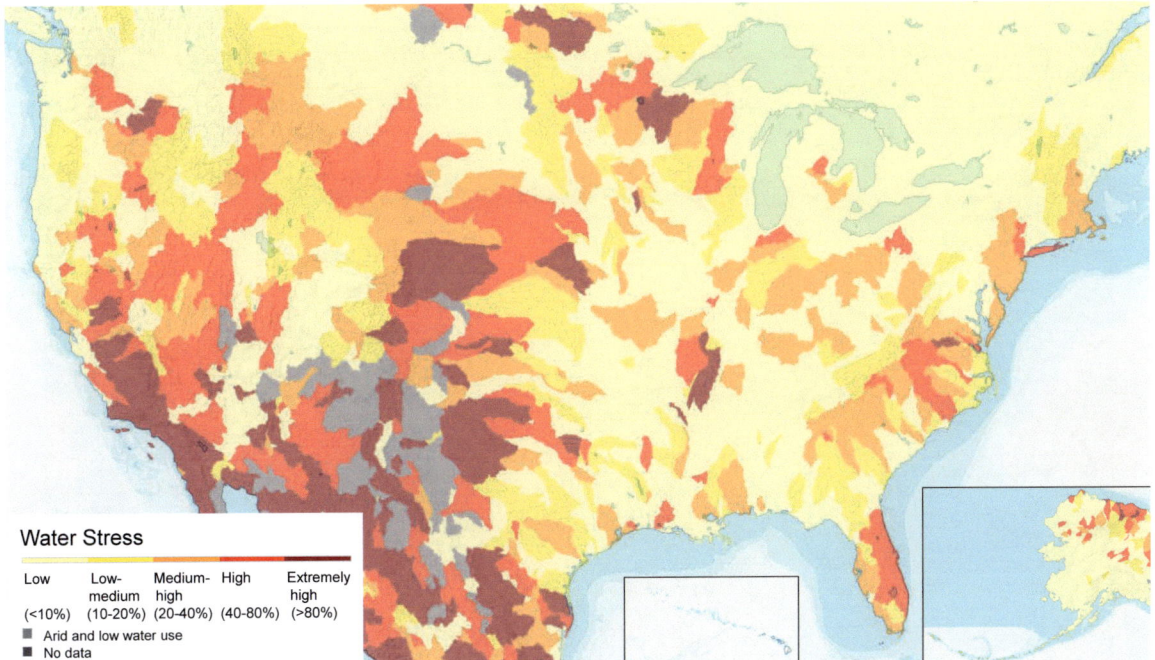

FIGURE 2-14 Risk of water stress across the United States as of 2019, where darker coloration indicates higher stress level.
SOURCE: World Resources Institute, 2019, "Aqueduct Water Risk Atlas," wri.org/aqueduct. CC BY 4.0.

2.4.4 Natural Gas

The U.S. natural gas infrastructure is well established and mature across the country (see Figure 2-15), with about 17,000 miles of gathering pipelines, 300,000 miles of transmission pipelines, and 2.3 million miles of distribution pipelines as of 2021 (PHMSA 2022a,b). When paired with CCS, natural gas can generate clean hydrogen and electricity. Natural gas usage is projected to decline globally in the coming decades in response to the transition to net-zero emissions, but the extent of this decline will depend on how much clean electricity and hydrogen is generated via natural gas with CCS versus via renewable or clean electricity. The United States has abundant natural gas reserves and infrastructure relative to future demand scenarios and is cost advantaged for production of clean hydrogen from natural gas and CCS.

As the country moves toward a future net-zero economy, natural gas will be gradually replaced by electricity or hydrogen for many end-use applications, potentially creating opportunities to repurpose the associated infrastructure. However, some scenarios, mostly depending on the energy price, anticipate the use of natural gas beyond mid-century, and low-carbon-intensity or renewable natural gas will likely remain energy sources for decades, presenting a coordination challenge during the transition. In any case, converting the remaining available natural gas pipelines to CO_2 trunk lines for capture and use or to hydrogen pipelines as a clean replacement for natural gas will be limited by metallurgical and/or pressure constraints, as discussed further in Sections 4.3.4 and 4.5.2. However, pipeline rights-of-way may be important for supportive infrastructure such as H_2 pipelines to enable clean transportation, energy storage, and industrial heating, as well as for manufacture of synthetic fuels and chemicals from CO_2.

Utilizing existing natural gas pipeline infrastructure for CO_2 transport provides a potential opportunity to address some of the myriad challenges associated with the buildout of pipelines. Pipeline projects are currently facing unprecedented pushback associated with societal and safety concerns (see, e.g., Payne 2021; PHMSA 2022d,e; Wittenberg 2022). These concerns need to be acknowledged and addressed through a transparent, inclusive process so as not to foster or perpetuate misinformation. The use of existing natural gas pipelines to transport CO_2 in gaseous form reduces this development risk, as it allows existing infrastructure to be used with fewer or no

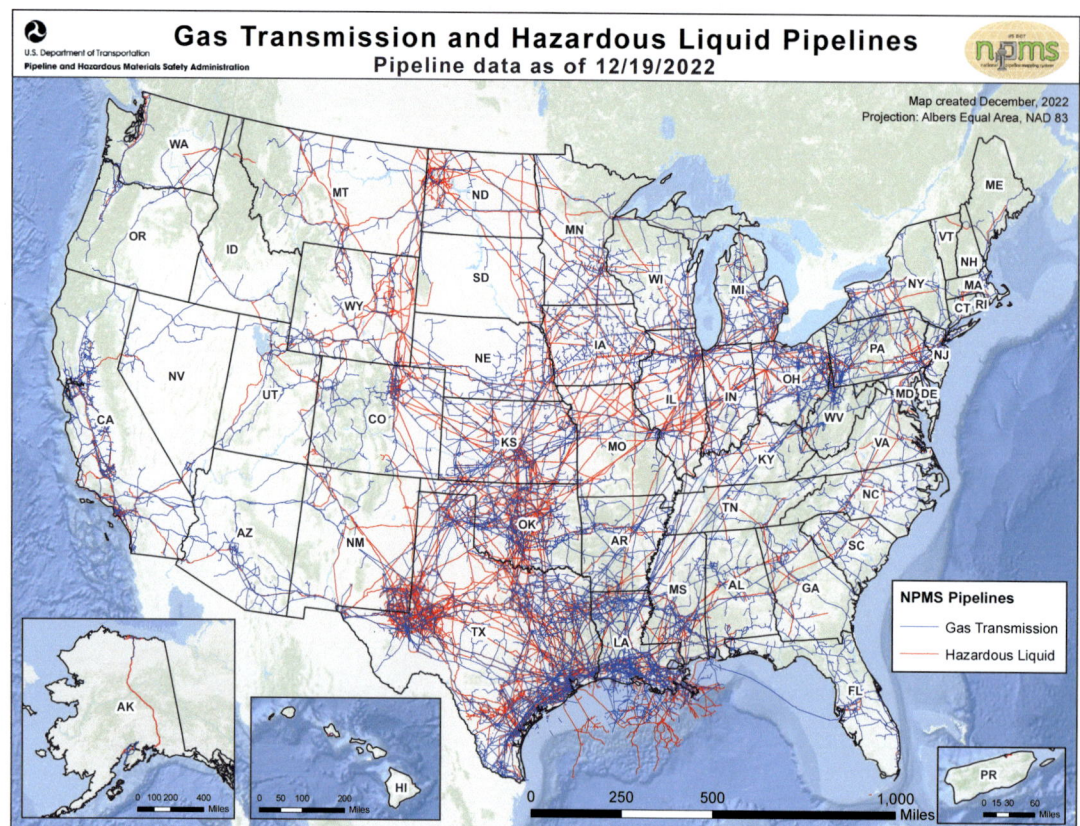

FIGURE 2-15 Map of gas transmission (blue) and hazardous liquid (red) pipelines in the United States as of March 2022. SOURCE: National Pipeline Mapping System, 2022, "Gas Transmission and Hazardous Liquid Pipelines." U.S. Department of Transportation Pipeline and Hazardous Materials Safety Administration, https://www.npms.phmsa.dot.gov/Documents/NPMS_Pipelines_Map.pdf.

added surface impacts. However, as discussed in Section 4.3, transporting CO_2 in the supercritical phase is much more economically viable than gas-phase transportation, and requires a higher transportation pressure than the safe operating pressure of natural gas pipelines. Therefore, achieving the scale of CCS or CCU necessary to meet climate objectives is unlikely to occur by relying solely on existing natural gas pipeline infrastructure. Section 4.3.4 provides more detail on considerations for retrofitting natural gas pipelines to transport CO_2.

2.5 FINDINGS ON EXISTING INFRASTRUCTURE FOR CO_2 UTILIZATION

FINDING 2.1 Barriers to Commercialization. As of 2022, few commercial processes transform CO_2 into useful products. Motivation to produce CO_2-derived products is limited because most products utilizing CO_2 are currently more expensive to produce than incumbents, especially fossil-fuel-derived products. Additionally, there are no requirements and limited global incentives in place to produce carbon-based products that can be certified as net-zero emissions. The availability of net-zero-carbon electricity, hydrogen, and heat is still limited, hindering their use in net-zero-emissions chemical and material production.

FINDING 2.2 Enhanced Oil Recovery. Enhanced oil recovery (EOR) currently accounts for the largest volume of CO_2 use in the United States, around 91,000 tons CO_2 per day as of 2020, primarily sourced from geologic

CO_2 reservoirs. The use of CO_2 for EOR is not a sustainable CO_2 utilization process, because it facilitates further fossil-fuel production and does not involve a chemical transformation of CO_2 into a product.

FINDING 2.3 Existing CO_2 Infrastructure. The majority of the existing infrastructure that could be used for carbon capture, utilization, and storage (CCUS) in the United States has been developed for EOR, and most CO_2 pipelines connect geologic or fossil sources of CO_2 with depleted oil reservoirs. These sources and sinks are generally not aligned with opportunities for sustainable CO_2 utilization in a net-zero emissions future, thereby limiting the opportunities to leverage existing infrastructure for development of CO_2 utilization processes.

FINDING 2.4 Co-location of CO_2 Capture and Use. Some CO_2 utilization processes can use flue gas directly (e.g., reactive capture) or be performed in situ at the point of CO_2 capture, both of which avoid the costs of CO_2 compression and transport. The feasibility of co-locating CO_2 capture and utilization will be influenced by the nature of the emissions source and the utilization process and product, economic viability, as well as the availability of other required low-carbon-emission energy sources (e.g., electricity and hydrogen) and infrastructure to transport the utilization product.

FINDING 2.5 Infrastructure for CO_2 Storage. To meet net-zero emissions goals, and given the much larger volumes of CO_2 that can be geologically stored versus used in products, most CO_2 captured from point sources, the atmosphere, and bodies of water will need to be sequestered in long-term geologic storage. Large-scale infrastructure for CCS will need to be developed in the coming decade, with some of this infrastructure potentially able to serve CO_2 utilization projects.

FINDING 2.6 Existing Hydrogen Infrastructure. The current U.S. infrastructure for hydrogen production and distribution—located almost entirely in the Gulf Coast region—supports a market for refining of petroleum into hydrocarbon fuels and petrochemicals that is anticipated to decline as the world transitions to net-zero emissions, as well as growing markets for hydrogen as a fuel, for use in ammonia production for fertilizer, and potentially as an energy carrier. The addition of carbon capture would allow the current hydrogen infrastructure for petroleum refining to be rededicated to future markets including sustainable CO_2 utilization products.

2.6 REFERENCES

Abramson, E., D. McFarlane, and J. Brown. 2020. *Transport Infrastructure for Carbon Capture and Storage: Whitepaper on Regional Infrastructure for Midcentury Decarbonization*. Minneapolis, MN: Great Plains Institute. https://www.betterenergy.org/wp-content/uploads/2020/06/GPI_RegionalCO2Whitepaper.pdf.

Abramson, E., E. Thomley, and D. McFarlane. 2022. *An Atlas of Carbon and Hydrogen Hubs for United States Decarbonization*. Minneapolis, MN: Great Plains Institute. https://scripts.betterenergy.org/CarbonCaptureReady/GPI_Carbon_and_Hydrogen_Hubs_Atlas.pdf.

Arcadia eFuels. 2022. "Arcadia eFuels Announces Its First eFuels Plant Location in Vordingborg, Denmark." https://www.arcadiaefuels.com/arcadia-efuels-announces-its-first-efuels-plant-location-in-vordingborg-denmark.

ARI (Advanced Resources International). 2021. "The U.S. CO_2 Enhanced Oil Recovery Survey." https://adv-res.com/pdf/ARI-2021-EOY-2020-CO2-EOR-Survey-OCT-21-2021.pdf.

Asahi Kasei Corporation. 2021. "Asahi Kasei Licenses Technology Package to Manufacture High-Purity Ethylene Carbonate and Dimethyl Carbonate Using CO_2 as Main Feedstock." https://www.asahi-kasei.com/news/2021/e210701.html.

Beswick, R.R., A.M. Oliveira, and Y. Yan. 2021. "Does the Green Hydrogen Economy Have a Water Problem?" *ACS Energy Letters* 6(9):3167–3169. https://doi.org/10.1021/acsenergylett.1c01375.

Bright, M. 2020. "Mapping the Progress and Potential of Carbon Capture, Use, and Storage." Third Way. June 1, 2020. https://www.thirdway.org/memo/mapping-the-progress-and-potential-of-carbon-capture-use-and-storage.

Bruhwiler, L., A.M. Michalak, R. Birdsey, J.B. Fisher, R.A. Houghton, D.N. Huntzinger, and J.B. Miller. 2018. "Chapter 1: Overview of the Global Carbon Cycle." Pp. 42–70 in *Second State of the Carbon Cycle Report. Sustained Assessment*, N. Cavallaro, G. Shrestha, R. Birdsey, M.A. Mayes, R.G. Najjar, S.C. Reed, P. Romero-Lankao, and Z. Zhu, eds. Washington, DC: U.S. Global Change Research Program. https://doi.org/10.7930/SOCCR2.2018.Ch1.

Burkart, M.D., N. Hazari, C.L. Tway, and E.L. Zeitler. 2019. "Opportunities and Challenges for Catalysis in Carbon Dioxide Utilization." *ACS Catalysis* 9(9):7937–7956. https://doi.org/10.1021/acscatal.9b02113.

Carbon Recycling International. n.d. "George Olah Renewable Methanol Plant: First Production of Fuel from CO_2 at Industrial Scale." https://www.carbonrecycling.is/project-goplant.

CEQ (Council on Environmental Quality). 2021. *Council on Environmental Quality Report to Congress on Carbon Capture, Utilization, and Sequestration*. Washington, DC. https://www.whitehouse.gov/wp-content/uploads/2021/06/CEQ-CCUS-Permitting-Report.pdf.

Circular Carbon Network. 2022. "Innovator Index." https://circularcarbon.org/innovator-index.

Constantz, B.R., M.A. Bewernitz, C.L. Camiré, S.-H. Kang, J. Schneider, and R.R. Wade. 2015. "Bioinspired Concrete." Pp. 297–308 in *Biotechnologies and Biomimetics for Civil Engineering*, F.P. Torgal, J.A. Labrincha, M.V. Diamanti, C.-P. Yu, and H.K. Lee, eds. Cham: Springer International Publishing. https://doi.org/10.1007/978-3-319-09287-4_13.

DAC (Direct Air Capture) Coalition. 2022. "The DAC Coalition: Technology Companies and Partners Working Together to Advance DAC Deployment." https://daccoalition.org/the-coalition.

Deutsch, T.G., S. Baker, P. Agbo, D.R. Kauffman, J. Vickers, and J.A. Schaidle. 2021. "Summary Report of the Reactive CO_2 Capture: Process Integration for the New Carbon Economy Workshop, February 18–19, 2020." Technical Report NREL/TP-5100-78466. Golden, CO: National Renewable Energy Laboratory. https://www.nrel.gov/docs/fy21osti/78466.pdf.

Dieter, C.A., M.A. Maupin, R.R. Caldwell, M.A. Harris, T.I. Ivahnenko, J.K. Lovelace, N.L. Barber, and K.S. Linsey. 2018. *Estimated Use of Water in the United States in 2015*. Water Availability and Use Science Program. Circular 1441 (supersedes USGS Open-File Report 2017-1131). Reston, VA: U.S. Geological Survey. https://pubs.usgs.gov/circ/1441/circ1441.pdf.

DiPietro, P., P. Balash, and M. Wallace. 2012. "A Note on Sources of CO_2 Supply for Enhanced-Oil-Recovery Operations." *SPE Economic & Management* 4(2):69–74.

DOE (U.S. Department of Energy). 2020. *Department of Energy Hydrogen Program Plan*. Washington, DC. https://www.hydrogen.energy.gov/pdfs/hydrogen-program-plan-2020.pdf.

DOE. 2021. "DOE Announces $12 Million for Direct Air Capture Technology." https://www.energy.gov/articles/doe-announces-12-million-direct-air-capture-technology.

DOE. 2022. "PEM Electrolyzer Capacity Installations in the United States." DOE Hydrogen Program Record 22001. https://www.hydrogen.energy.gov/pdfs/22001-electrolyzers-installed-in-united-states.pdf.

DOE-EERE (Office of Energy Efficiency & Renewable Energy). n.d. "Hydrogen Pipelines." https://www.energy.gov/eere/fuelcells/hydrogen-pipelines.

DOE-FECM (Office of Fossil Energy and Carbon Management). 2022. "Selections for Funding Opportunity Announcement 2560: Direct Air Capture Combined with Dedicated Long-Term Carbon Storage, Coupled to Existing Low-Carbon Energy." https://www.energy.gov/fecm/articles/selections-funding-opportunity-announcement-2560-direct-air-capture-combined.

DOE-HFTO (Hydrogen and Fuel Cell Technologies Office). n.d. "DOE Technical Targets for Hydrogen Delivery." https://www.energy.gov/eere/fuelcells/doe-technical-targets-hydrogen-delivery.

Donlan, M., and C. Trabucchi. 2011. "Valuation of Consequences Arising from CO_2 Migration at Candidate CCS Sites in the US." *Energy Procedia* 4:2222–2229. https://doi.org/10.1016/j.egypro.2011.02.110.

Dowling, J.A., K.Z. Rinaldi, T.H. Ruggles, S.J. Davis, M. Yuan, F. Tong, N.S. Lewis, and K. Caldeira. 2020. "Role of Long-Duration Energy Storage in Variable Renewable Electricity Systems." *Joule* 4(9):1907–1928. https://doi.org/10.1016/j.joule.2020.07.007.

Dvory, N.Z., and M.D. Zoback. 2021." Prior Oil and Gas Production Can Limit the Occurrence of Injection-Induced Seismicity: A Case Study in the Delaware Basin of Western Texas and Southeastern New Mexico, USA." *Geology* 49(10):1198–1203. https://doi.org/10.1130/G49015.1.

Edwards, R.W.J., and M.A. Celia. "Infrastructure to Enable Deployment of Carbon Capture, Utilization, and Storage in the United States." *Proceedings of the National Academy of Sciences* 115(38):E8815–E8824. https://doi.org/10.1073/pnas.1806504115.

EFI (Energy Futures Initiative). 2021. *The Future of Clean Hydrogen in the United States: Views from Industry, Market Innovators, and Investors*. Part of the EFI Report Series *From Kilograms to Gigatons: Pathways for Hydrogen Market Formation in the United States*. Washington, DC. https://energyfuturesinitiative.org/wp-content/uploads/sites/2/2022/03/The-Future-of-Clean-Hydrogen-in-the-U.S._Report-1.pdf.

EIA (U.S. Energy Information Administration). 2021. "What Are Hydrocarbon Gas Liquids?" Updated October 26. https://www.eia.gov/energyexplained/hydrocarbon-gas-liquids.

EIA. 2022. "What Is U.S. Electricity Generation by Energy Source?" Frequently Asked Questions (FAQS). https://www.eia.gov/tools/faqs/faq.php?id=427&t=3.

EPA (U.S. Environmental Protection Agency). 2017. "Greenhouse Gas Reporting Program Dataset." FLIGHT database. https://www.epa.gov/ghgreporting/data-sets.

EPA. 2022. *Inventory of U.S. Greenhouse Gas Emissions and Sinks: 1990–2020*. EPA 430-R-22-003. https://www.epa.gov/system/files/documents/2022-04/us-ghg-inventory-2022-main-text.pdf.

FCHEA (Fuel Cell and Hydrogen Energy Association). 2020. "Road Map to a US Hydrogen Economy." https://www.fchea.org/us-hydrogen-study.

Global CCS Institute. 2021. *Blue Hydrogen*. Washington, DC. https://www.globalccsinstitute.com/wp-content/uploads/2021/04/Circular-Carbon-Economy-series-Blue-Hydrogen.pdf.

Global CCS Institute. 2022. "CO_2RE Facilities Database." https://co2re.co/FacilityData.

Global CO_2 Initiative. 2018. "Global Roadmap Study of CO_2U Technologies." Technical Report. University of Michigan. https://hdl.handle.net/2027.42/146529.

Grand View Research. 2022. *Carbon Dioxide Market Size, Share & Trends Analysis Report by Source (Ethyl Alcohol, Ethylene Oxide), by Application (Food & Beverages, Oil & Gas, Medical), by Region, and Segment Forecasts, 2022–2030*. San Francisco. https://www.grandviewresearch.com/industry-analysis/carbon-dioxide-market.

Greenwood, A. 2021. "CO_2-Based Methanol from US Celanese JV Resemble Natgas-Based Costs." Independent Commodity Intelligence Services. https://www.icis.com/explore/resources/news/2021/04/23/10631870/co2-based-methanol-from-us-celanese-jv-resemble-natgas-based-costs.

Grubert, E., and K.T. Sanders. 2018. "Water Use in the United States Energy System: A National Assessment and Unit Process Inventory of Water Consumption and Withdrawals." *Environmental Science & Technology* 52(11):6695–6703. https://doi.org/10.1021/acs.est.8b00139.

Haldor-Topsoe. 2022. "G2L™ eFuels Technology." Topsoe. https://www.topsoe.com/our-resources/knowledge/our-products/process-licensing/g2ltm-efuels-technology?hsLang=en.

Hausfather, Z. 2021. "Global CO_2 Emissions Have Been Flat for a Decade, New Data Reveals." *CarbonBrief*, November 4. https://www.carbonbrief.org/global-co2-emissions-have-been-flat-for-a-decade-new-data-reveals.

Hills, C.D., N. Tripathi, R.S. Singh, P.J. Carey, and F. Lowry. 2020. "Valorisation of Agricultural Biomass-Ash with CO_2." *Scientific Reports* 10(1):13801. https://doi.org/10.1038/s41598-020-70504-1.

Hobson, C. 2018. *Renewable Methanol Report*. Madrid, Spain: Methanol Institute. https://www.methanol.org/wp-content/uploads/2019/01/MethanolReport.pdf.

Holmes, K.J., E. Zeitler, M. Kerxhalli-Kleinfield, and R. DeBoer. 2021. "Scaling Deep Decarbonization Technologies." *Earth's Future* 9(11):e2021EF002399. https://doi.org/10.1029/2021EF002399.

IDEALHy (Integrated Design for Efficient Advanced Liquefaction of Hydrogen). n.d. "Liquid Hydrogen Outline." https://www.idealhy.eu/index.php?page=lh2_outline.

IEA (International Energy Agency). 2019. *Putting CO_2 to Use: Creating Value from Emissions*. Paris. https://iea.blob.core.windows.net/assets/50652405-26db-4c41-82dc-c23657893059/Putting_CO2_to_Use.pdf.

IEA. 2021. *Global Energy Review 2021*. Paris: IEA. https://www.iea.org/reports/global-energy-review-2021.

IHS Markit. 2021. "Carbon Dioxide Market Research: Carbon Dioxide (CO_2) Outlook, Supply & Demand, Forecast and Analysis." In *Chemical Economics Handbook—Carbon Dioxide*. S&P Global. https://ihsmarkit.com/products/carbon-dioxide-chemical-economics-handbook.html.

Ineratec. 2017. "Sustainable E-fuels for Aviation." https://ineratec.de/en/e-fuels-for-aviation.

Jones, A.C., and A.J. Lawson. 2021. *Carbon Capture and Sequestration in the United States*. Washington, DC: Congressional Research Service. https://sgp.fas.org/crs/misc/R44902.pdf.

Kahler, F., M. Carus, O. Porc, and C. vom Berg. 2021. *Turning Off the Tap for Fossil Carbon—Future Prospects for a Global Chemical and Derived Material Sector Based on Renewable Carbon*. Hurth, Germany: Nova Institute for Ecology and Innovation. https://renewable-carbon.eu/publications/product/turning-off-the-tap-for-fossil-carbon-future-prospects-for-a-global-chemical-and-derived-material-sector-based-on-renewable-carbon.

Kelly, S., and T. Polansek. 2020. "Side Effects: Fuel Demand Crash Shuts U.S. Ethanol Plants, Meatpackers Lack Refrigerant." *Reuters*, April 7. https://www.reuters.com/article/us-health-coronavirus-carbon-dioxide/side-effects-fuel-demand-crash-shuts-u-s-ethanol-plants-meatpackers-lack-refrigerant-idUSKBN21P2DK.

Küngas, R., P. Blennow, T. Heiredal-Clausen, T. Holt, J. Rass-Hansen, S. Primdahl, and J. Bøgild Hansen. 2017. "ECOs—A Commercial CO_2 Electrolysis System Developed by Haldor Topsoe." *ECS Transactions* 78(1):2879. https://doi.org/10.1149/07801.2879ecst.

Larson, E., C. Greig, J. Jenkins, E. Mayfield, A. Pascale, C. Zhang, J. Drossman, et al. 2021. *Net-Zero America: Potential Pathways, Infrastructure, and Impacts*. Final Report. Princeton, NJ: Princeton University. https://netzeroamerica.princeton.edu/the-report.

LEP (Labor Energy Partnership). 2021. *Building to Net-Zero: A U.S. Policy Blueprint for Gigaton-Scale CO_2 Transport and Storage Infrastructure*. Washington, DC: Energy Futures Initiative and AFL-CIO. https://laborenergy.org/wp-content/uploads/2021/10/LEP-Building_to_Net-Zero-June-2021-v4.pdf.

Madigan, J. 2020. "Carbon Dioxide Production." US Industry (Specialized) Report OD4929. IBISWorld.

Mitsubishi Gas Chemical Company, Inc. 2021. "Mitsubishi Gas Chemical to Launch 'Circular Carbon Methanol' Production." https://www.mgc.co.jp/eng/corporate/news/files/210330e.pdf.

Monkman, S., and M. MacDonald. 2017. "On Carbon Dioxide Utilization as a Means to Improve the Sustainability of Ready-Mixed Concrete." *Journal of Cleaner Production* 167(November):365–375. https://doi.org/10.1016/j.jclepro.2017.08.194.

Murphy, C., T. Mai, Y. Sun, P. Jadun, M. Muratori, B. Nelson, and R. Jones. 2021. *Electrification Futures Study: Scenarios of Power System Evolution and Infrastructure Development for the United States*. NREL/TP-6A20-72330. Golden, CO: National Renewable Energy Laboratory. https://doi.org/10.2172/1762438.

NASEM (National Academies of Sciences, Engineering, and Medicine). 2022. *A Research Strategy for Ocean-Based Carbon Dioxide Removal and Sequestration*. Washington, DC: The National Academies Press. https://doi.org/10.17226/26278.

National Carbon Capture Center. 2022. "Homepage." https://www.nationalcarboncapturecenter.com.

NETL (National Energy Technology Laboratory). 2017a. *Best Practices: Monitoring, Verification, and Accounting (MVA) for Geologic Storage Projects*. DOE/NETL-2017/1847, Revised Edition. U.S. Department of Energy. https://netl.doe.gov/sites/default/files/2018-10/BPM-MVA-2012.pdf.

NETL. 2017b. "FE/NETL CO_2 Saline Storage Cost Model (2017)." U.S. Department of Energy. Last update, September (version 3). https://www.netl.doe.gov/research/energy-analysis/search-publications/vuedetails?id=2403.

Northern Lights. 2021. "What It Takes to Ship CO_2." https://northernlightsccs.com/news/what-it-takes-to-ship-co2.

NPC (National Petroleum Council). 2021a. "CO_2 Geologic Storage." Chapter 7 in *Meeting the Dual Challenge: A Roadmap to At-Scale Deployment of Carbon Capture, Use, and Storage, 2021*. A Report of the National Petroleum Council. https://dualchallenge.npc.org/files/CCUS-Chap_7-030521.pdf.

NPC. 2021b. "CO_2 Transport." Chapter 6 in *Meeting the Dual Challenge: A Roadmap to At-Scale Deployment of Carbon Capture, Use, and Storage. 2021*. A Report of the National Petroleum Council. https://dualchallenge.npc.org/files/CCUS-Chap_6-030521.pdf.

NPMS (National Pipeline Mapping System). 2022. "Gas Transmission and Hazardous Liquid Pipelines." U.S. Department of Transportation Pipeline and Hazardous Materials Safety Administration. https://www.npms.phmsa.dot.gov/Documents/NPMS_Pipelines_MAP.pdf.

Omodolor, I.S., H.O. Otor, J.A. Andonegui, B.J. Allen, and A.C. Alba-Rubio. 2020. "Dual-Function Materials for CO_2 Capture and Conversion: A Review." *Industrial & Engineering Chemistry Research* 59(40):17612–176331. https://doi.org/10.1021/acs.iecr.0c02218.

PARC (Palo Alto Research Center). 2022. "Commercial Methane Pyrolysis Efforts." https://docs.google.com/spreadsheets/d/1IcMP7WlmhntRz3hKvVjvr2IwrFprgCe-1bYAtY56eOk/edit#gid=0.

Payne, K. 2021. "Proposed Carbon Dioxide Pipeline Draws Opposition from Iowa Farmers and Environmentalists Alike." Iowa Public Radio. October 13. https://www.iowapublicradio.org/ipr-news/2021-10-13/proposed-carbon-dioxide-pipeline-draws-opposition-from-iowa-farmers-and-environmentalists-alike.

PHMSA (Pipeline and Hazardous Materials Safety Administration). 2022a. "Annual Report Mileage for Gas Transmission & Gathering Systems." U.S. Department of Transportation. June 1. https://www.phmsa.dot.gov/data-and-statistics/pipeline/annual-report-mileage-natural-gas-transmission-gathering-systems.

PHMSA. 2022b. "Annual Report Mileage for Gas Distribution Systems." U.S. Department of Transportation. June 1. https://www.phmsa.dot.gov/data-and-statistics/pipeline/annual-report-mileage-gas-distribution-systems.

PHMSA. 2022c. "Annual Report Mileage for Hazardous Liquid or Carbon Dioxide Systems." U.S. Department of Transportation. May 2. https://www.phmsa.dot.gov/data-and-statistics/pipeline/annual-report-mileage-hazardous-liquid-or-carbon-dioxide-systems.

PHMSA. 2022d. *Failure Investigation Report—Denbury Gulf Coast Pipelines, LLC–Pipeline Rupture/Natural Force Damage*. Washington, DC: U.S. Department of Transportation. https://www.phmsa.dot.gov/sites/phmsa.dot.gov/files/2022-05/Failure%20Investigation%20Report%20-%20Denbury%20Gulf%20Coast%20Pipeline.pdf.

PHMSA. 2022e. "PHMSA Announces New Safety Measures to Protect Americans from Carbon Dioxide Pipeline Failures After Satartia, MS Leak." U.S. Department of Transportation. May 26. https://www.phmsa.dot.gov/news/phmsa-announces-new-safety-measures-protect-americans-carbon-dioxide-pipeline-failures.

Pires da Mata Costa, L., D.M. Vaz de Miranda, A.C. Couto de Oliveira, L. Falcon, M.S. Silva Pimenta, I. G. Bessa, S. Juarez Wouters, M.H.S. Andrade, and J.C. Pinto. 2021. "Capture and Reuse of Carbon Dioxide (CO_2) for a Plastics Circular Economy: A Review." *Processes* 9(5):759. https://doi.org/10.3390/pr9050759.

Rubin, E.S., C. Short, G. Booras, J. Davison, C. Ekstrom, M. Matuszewski, and S. McCoy. 2013. "A Proposed Methodology for CO_2 Capture and Storage Cost Estimates." *International Journal of Greenhouse Gas Control* 17(September):488–503. https://doi.org/10.1016/j.ijggc.2013.06.004.

Ruth, M.F., P. Jadun, N. Gilroy, E. Connelly, R. Boardman, A.J. Simon, A. Elgowainy, and J. Zuboy. 2020. *The Technical and Economic Potential of the H_2@Scale Concept Within the United States*. NREL/TP-6A20-77610. Golden, CO: National Renewable Energy Laboratory. https://www.nrel.gov/docs/fy21osti/77610.pdf.

Scott, A. 2019. "Avantium Is Making Bioethylene Glycol." *C&EN Global Enterprise* 97(45):14. https://doi.org/10.1021/cen-09745-buscon10.

Shaner, M.R., S.J. Davis, N.S. Lewis, and K. Caldeira. 2018. "Geophysical Constraints on the Reliability of Solar and Wind Power in the United States." *Energy & Environmental Science* 11(4):914–925. https://doi.org/10.1039/C7EE03029K.

Shao, B., Y. Zhang, Z. Sun, J. Li, Z. Gao, Z. Xie, J. Hu, and H. Liu. 2022. "CO_2 Capture and In-Situ Conversion: Recent Progresses and Perspectives." *Green Chemical Engineering* 3(3):189–198. https://doi.org/10.1016/j.gce.2021.11.009.

Solidia (Solidia Technologies). 2022. "The U.S. Department of Energy Grants $2.1M to Solidia Technologies to Develop CO2 Capture and Utilization Technologies for Building Materials." *3BL CSRwire* May 23. https://www.csrwire.com/press_releases/745286-us-department-energy-grants-21m-solidia-technologies-develop-co2-capture-and.

Wittenberg, A. 2022. "Strange Bedfellows: Farmers and Big Greens Square Off Against Biden and the GOP." *Politico*, May 29. https://www.politico.com/news/2022/05/29/iowa-manchin-carbon-capture-pipeline-00030361.

World Energy Council. 2021. *Hydrogen on the Horizon: Ready, Almost Set, Go? Innovation Insights Briefing*. London: World Energy Council. https://www.worldenergy.org/assets/downloads/Innovation_Insights_Briefing_-_Hydrogen_on_the_Horizon_-_Ready%2C_Almost_Set%2C_Go_-_July_2021.pdf.

Wyoming ITC (Integrated Test Center). n.d. "Homepage." https://www.wyomingitc.org.

XPRIZE. 2022. "XPRIZE and the Musk Foundation Award $15M to Prize Milestone." https://www.xprize.org/prizes/elonmusk/articles/xprize-and-the-musk-foundation-award-15m-to-prize-milestone-winners-in-100m-carbon-removal-competition.

XPRIZE. n.d. "Turning CO_2 into Products: NRG COSIA Carbon XPRIZE Guidelines." https://www.xprize.org/prizes/carbon/guidelines.

Yang, Q., X. Liu, S. Zhu, W. Huang, and D. Zhang. 2019. "Efficient Utilization of CO2 in a Coal to Ethylene Glycol Process Integrated with Dry/Steam-Mixed Reforming: Conceptual Design and Technoeconomic Analysis." *ACS Sustainable Chemistry & Engineering* 7(3):3496–3510. https://doi.org/10.1021/acssuschemeng.8b05757.

ZEP (Zero Emissions Platform). 2011. *The Costs of CO_2 Transport: Post-Demonstration CCS in the EU*. Brussels, Belgium: European Technology Platform for Zero Emission Fossil Fuel Power Plants. https://www.globalccsinstitute.com/archive/hub/publications/119811/costs-co2-transport-post-demonstration-ccs-eu.pdf.

3

Potential Uses of CO_2 in Commercial Products

3.1 FRAMING, INTRODUCTION, AND SCOPE OF CHAPTER

Carbon is the central element of many manufactured products, notably plastics, fuels, and commodity chemicals; however, current production relies predominantly on fossil sources of carbon, carbon-emitting inputs such as fossil-fuel-powered heat, electricity, and transportation, and fossil-derived co-reagents such as hydrogen. These processes and, upon degradation, the embodied carbon in the products, result in greenhouse gas (GHG) emissions to the atmosphere. Emissions from chemicals and materials production and embodied carbon in products will need to be reduced to net zero for long-term sustainability. One route to net-zero emissions production of carbon-based chemicals and materials is to replace fossil sources of carbon with non-fossil-derived carbon dioxide (CO_2), and reduce to zero the emissions associated with all other inputs to the utilization process.

In the context of this report, it is useful to categorize products made from CO_2 into two tracks that are distinguished by the lifetime of the products: Track 1 with lifetime greater than 100 years and Track 2 with lifetime less than 100 years (Sick et al. 2022a). Although the exact lifetime to distinguish between the two tracks is debatable, 100 years is used here following a practice promoted by the United Nations Framework Convention on Climate Change (UNFCCC n.d.). The climate relevance of products in the tracks will be different. Track 1 products, especially concrete and aggregates, are considered durable storage options for CO_2. Likewise, most polymers have lifetimes of hundreds of years and therefore can hold carbon long enough to provide a removal-related climate benefit, if appropriately treated to prevent GHG emissions through degradation or incineration. Track 2 products, on the other hand, decompose back into CO_2 over a shorter timescale. Thus, their primary climate benefit is staked on avoiding the use of fossil carbon in their production (see Figure 3-1). CO_2 utilization can replace fossil carbon embedded in chemicals, fuels, and materials with renewable carbon from CO_2. The embedded carbon in chemicals and fuels will be released to the atmosphere upon degradation of chemicals and materials over short or long timescales, representing 450 million metric tons (MMT) carbon per year globally for chemicals and derived materials (Kahler et al. 2021), and 3,337 and 2,062 MMT carbon in oil and gas fuels, respectively, as of 2019 (Friedlingstein et al. 2022; Global Carbon Project 2021).[1] In addition to embodied carbon, emissions associated with chemical and fuel production, along with other aspects of the product life cycle, are significant. Net-zero carbon use for chemicals and fuels can be feasible, provided that all inputs sum to a zero-carbon budget on a life cycle basis.

---

[1] Embodied carbon in oil and gas fuels was estimated using the estimated emissions from combustion of such fuels in 2019.

The range of product categories accessible from CO_2 feedstock is very large. A 2019 National Academies study of CO_2 utilization status and research needs summarized the scope of technologies and products available for mineral, chemical, and biological processes (NASEM 2019). Mineralization processes were deemed the most highly developed potential growth areas for utilization, including carbonated precast and ready-mix concrete and aggregate materials. Chemical and biological processes are available to produce various chemicals and materials directly from CO_2, and prospects under development include one-carbon (C1), two-carbon (C2), and multi-carbon (C2+) hydrocarbons and oxygenates, monomers and polymers, and other materials. A 2016 study examined a similar scope of possible products and projected the potential annual use of CO_2 to make these products by 2030 to be between 2 and 8 gigatonnes (Gt), generating an annual revenue stream between $0.5 trillion and $2 trillion (Global CO_2 Initiative 2018). Subsequent studies reported similar magnitudes of CO_2 consumption and future market opportunity for CO_2 utilization (Biniek et al. 2020; CSIRO 2022; Hepburn et al. 2019; NASEM 2019; Sick et al. 2022b; some of which are summarized in Table 3-1). At this time, the future scope of this emerging industry is highly uncertain and thus estimates of potential will vary widely. Local needs, resource availability, and regulatory and incentive frameworks also will lead to significantly different outcomes for different regions. In general, the role of CO_2 utilization is to provide a sustainable source of carbon that can enable net-zero production of chemicals and materials, rather than contribute significantly to durable emissions mitigation (see, e.g., Mac Dowell et al. 2017).

CO_2 utilization enables net-zero or net-negative chemicals and materials production and provides opportunities to design new products and processes with preferred attributes. In some cases, especially for Track 2 chemicals, CO_2 substitutes for fossil carbon to produce the same product or products of similar function. Even in those cases, CO_2 offers opportunities to re-design products and processes, especially taking advantage of the

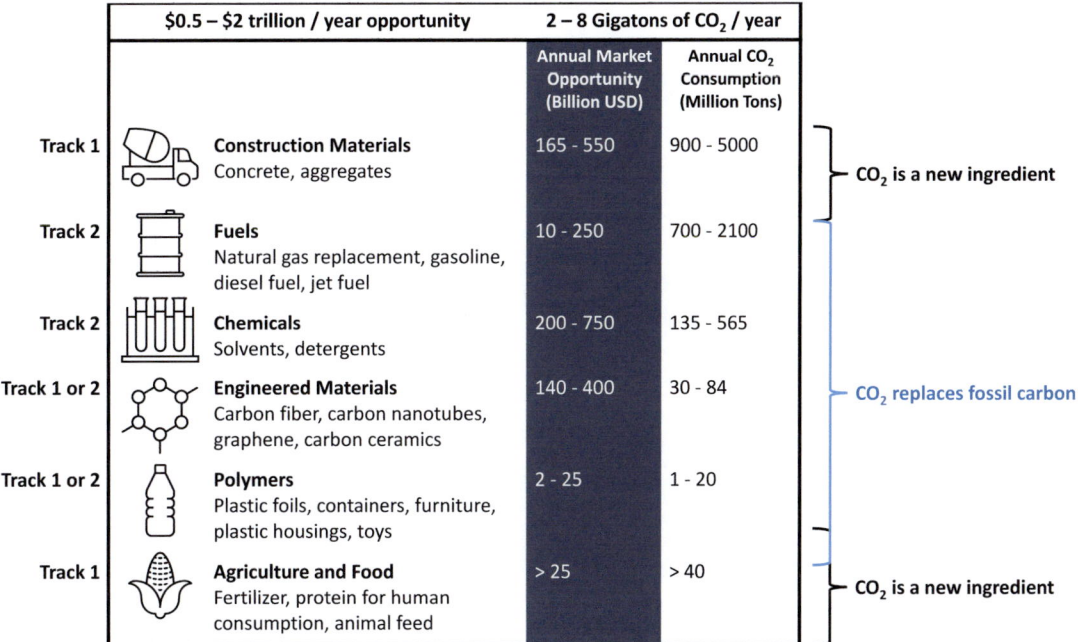

FIGURE 3-1 Estimated annual CO_2 utilization and revenue potential by 2050. Note that animal feed specifically refers to animal feed that can be produced from utilization of CO_2 by microbes, and not to all sources of animal feed, many of which are out of scope of this report.
SOURCE: Committee generated based on data from V. Sick, G. Stokes, F. Mason, et al., 2022, *Implementing CO_2 Capture and Utilization at Scale and Speed: The Path to Achieving Its Potential*, Technical Report, Global CO_2 Initiative, University of Michigan, https://dx.doi.org/10.7302/5825. Reproduced with permission, Global CO_2 Initiative. CC BY 4.0.

TABLE 3-1 Summary of Estimated and Projected Market Value and Volume of Products and CO_2 Amounts Utilized for Various Products

CO_2 Utilization Market	Product Market Size	
Specific Product Example or Market Cluster	2015 Total Market Size and 2030 Expected Total Market Size[a]	2030 Potential Market for Product Covered by Utilized Carbon[a]
Building Materials		
Concrete and concrete aggregates	2015: 20–30 Gt 2030 projection: 40 Gt	6.5–16.5 Gt
Carbonated aggregates	2015: 25–35 Gt 2030 projection: 50 Gt	1–10.5 Gt
Fuels		
Liquid fuels	2015: 800 billion to 1 trillion gallons 2030 projection: 1 trillion+ gallons	7–165 billion gallons
Jet fuel		
Methane	2015: 3–4 trillion cubic meters 2030 projection: over 4–5 trillion cubic meters	4–165 billion cubic meters
Chemical Intermediates		
All chemicals		
Methanol	2015: 60–70 Mt 2030 projection: 190 Mt	4–34 Mt (for fuels), 1.3–9.3 Mt (for chemical intermediates)
Syngas	2015: 130–150 GW 2030 projection: 500 GW	15–265 GW
Formic acid	2015: 0.5–0.7 Mt 2030 projection: 1 Mt	10–475 kt
Polymers and Materials		
Polyurethane		
Polyhydroxyalkanoates		
Polyols and polycarbonates	2015: 8–10 Mt 2030 projection: 17 Mt	0.4–6.8 Mt
Carbon fiber		
Carbon black		
Carbon nanotubes		

Utilization of CO$_2$ in Products (Mt CO$_2$/year)		Market Value ($)
2030 Potential CO$_2$ Utilization in Product[b]	2050 Potential CO$_2$ Utilization in Product	Market Value of Product in 2030, 2035, 2040,[e] and 2050[d]
150	100–1,400[c] 24–1,300 (precast concrete)[d] 1,000–9,500[d]	2030: $50 billion 2035: $150 billion 2040: $450 billion 2050: $182–$337 billion (aggregates), $623–$666 billion (precast concrete)
15	1,000–4,200[c] 14–10,200[d]	2030: $17 billion 2035: $25 billion 2040: $36 billion 2050: $5–$1,849 billion
	260–4,400[d]	2050: $16–$214 billion
	300–600[c] 260–580[d]	2030: $300 million 2035: $3 billion 2040: $25 billion 2050: $96–$183 billion
	0.5–1[d]	2030: $64 million 2035: $200 million 2040: $300 million 2050: $700–$840 million
10 (plastics, including polyurethane)	1.8–13	2030: $13 billion 2035: $24 billion 2040: $40 billion 2050: $130–$191 billion
		2030: $10 million 2035: $30 million 2040: $60 million
0.1		
	40–200	2050: $14–$66 billion
		2030: $3 million 2035: $21 million 2040: $32 million

continued

TABLE 3-1 Continued

CO_2 Utilization Market	Product Market Size
Algae and Other Microorganisms	
Microalgae (processed to create biofuels or food additives)	
Fish meal protein alternative	

NOTES: This table represents a collection of available data for several broad product volume and market value studies. Absences of values indicate that the studies did not examine or make estimates for the categories. The volumes and values across different studies are independent estimates with different assumptions and are not directly comparable.
SOURCES: [a] Global CO_2 Initiative (2018); [b] Biniek et al. (2020); [c] Hepburn et al. (2019); [d] Sick et al. (2022b); [e] and Lux Research (2020).

oxygen content of CO_2 in the production of oxygenated products like polycarbonates (Zimmerman et al. 2020). For others, the introduction of carbon into these products is new, namely construction materials in Track 1 such as CO_2-cured concrete or carbonated aggregates. In addition to creating large-scale CO_2 removal opportunities, mineral carbonation to form building materials offers the chance to develop entirely new materials with new properties and performance, though these may trigger the need for extensive documentation and updated codes and standards.

Maximum growth rates of CO_2 utilization estimated in market studies have not been achieved to date, due to the high cost of CO_2 and zero-emission energy versus fossil inputs, a lack of supportive legislation and policy to incentivize production and use of net-zero carbon products, a lack of infrastructure for CO_2 and enabling inputs, and frequently a perception of immaturity of underlying technologies requiring a long duration to profitability.

Even with a supportive environment for producing net-zero carbon products, CO_2 utilization will compete with other means of reducing the lifecycle emissions of products and materials, including switching from hydrocarbon fuels to electricity and hydrogen and replacing fossil-carbon feedstocks with bio-based or recycled carbon inputs. Replacing hydrocarbon fuels with hydrogen or electricity rather than CO_2-based synthetic fuels, when feasible, requires less energy. The energy cost needed to convert CO_2 into a fuel can be two to four times the amount required to use electricity or clean hydrogen directly as an energy carrier (Bazzanella and Ausfelder 2017; Herzog 2018). Therefore, the production of hydrocarbon fuels may be motivated primarily by use cases where electricity or/and hydrogen are not (currently) a feasible option, for example, long-haul aviation. On the other hand, many carbon-based chemicals and materials cannot be substituted readily with noncarbon alternatives. Where carbon-based materials are required, fossil-free production from CO_2 (Kätelhön et al. 2019) will compete with alternative sources of carbon such as biomass and recycled products (Berg et al. 2020). Biomass feedstock, especially waste biomass as a lower-cost competitor to CO_2, is likely to be an important source of carbon in the future, but there is limited available land to derive carbon feedstocks from plants, particularly if waste biomass is being used. Recycled products are also an important carbon feedstock competing with CO_2, but even as challenges in chemical recycling are overcome, some processes will favor CO_2 as a starting material, and in some instances, it will be impractical to collect all product material for recycling. In addition, the demand for carbon-based products and associated export opportunities will continue to increase globally, as economies grow to meet the needs of billions of currently underserved people.

Development and deployment of CO_2-derived product technologies needs to be accompanied and guided by rigorous life cycle assessments (LCAs), coupled with techno-economic assessments (TEAs), to evaluate

Utilization of CO_2 in Products (Mt CO_2/year)	Market Value ($)
200–900	
	2030: $25 million
	2035: $3 billion
	2040: $13 billion
	2050: $18–$921 billion

benefits and risks to the environment as well as economic viability. LCAs and TEAs can be performed at different levels of detail appropriate to the project scope and stage, and the outcomes of such studies are inputs for decision making along the entire path from invention to full-scale commercialization. A recent international community effort developed CO_2-specific guidelines for LCAs and TEAs (Zimmermann et al. 2018), and a web resource initiated and supported by the U.S. Department of Energy (DOE) was made available in 2021 (Global CO_2 Initiative 2022a). The AssessCCUS resource includes databases of information as well as tools for conducting TEAs and LCAs. This effort begins to address 2022 guidance from the White House Council on Environmental Quality that called for consolidating and publishing a repository for LCA methodology, results, and information related to CO_2 capture and utilization (CCU) and carbon dioxide removal (CDR) (CEQ, 2022).

Choosing research directions, determining priority criteria for environmental risks, or directing policy support involves multifactor decision processes of potentially high complexity and ambiguity (Cremonese et al. 2022). For CO_2 utilization, comprehensive analyses and scenario studies are needed to identify technologies and processes that produce needed chemicals, offer environmental improvements, and have the best prospects of being cost competitive against other options for meeting the same goal (Fernández-Dacosta et al. 2017). Trade-offs need to be evaluated using a suitable unit of analysis that allows results to be compared. This applies both to comparing different CO_2 utilization production pathways and to comparing CO_2 utilization pathways with those that use other sustainable carbon feedstocks (Artz et al. 2018; Ravikumar et al. 2021b). Equally or more complex are comparisons between different products, such as concrete and chemicals, to identify the largest impact on CO_2 emissions (Ravikumar et al. 2021a), or different ways of meeting a greater goal, such as net-zero emissions to the atmosphere.

The remainder of this chapter describes in more detail the products and processes relevant for CO_2 utilization in a net-zero emissions future, in order to inform needs for infrastructure and policies described later in the report. The potential climate impact of net-zero CO_2 utilization processes and products is assessed, with a focus on either durable products that can offer long-term storage of CO_2 or nondurable products that will represent a large-scale displacement of fossil-based chemicals in a net-zero carbon economy. The chapter also enumerates needs for additional inputs such as energy, hydrogen, water, land, and facilities. As noted in Chapter 1, this report focuses on chemically transformative uses of CO_2 for a net-zero carbon future and does not include an analysis of CO_2 use for enhanced oil and gas recovery (EOR, EGR), as a process fluid (cooling, cutting, fire extinction), or for highly transient products (e.g., beverages, supercritical fluid extractions, low-temperature drying, insulation).

3.2 FUTURE SOURCES OF CO_2 FOR UTILIZATION

Chapter 1 describes the need to reduce atmospheric concentrations of GHGs; the most likely path to reduce emissions is to decarbonize the energy system via a shift from fossil-fuel combustion to zero-emissions energy conversion and storage. Decarbonization of the energy system means that most sources of CO_2 emissions today (see Section 2.1) are likely to be eliminated or significantly reduced by 2050. The sources that remain in the long term may be those that are most technically difficult or expensive to eliminate, including CO_2 emissions from the combustion of fuels for heavy-duty transportation such as aviation and shipping, and industrial process emissions such as those from steel or cement manufacturing, among others. Some of these emissions may be from combustion of fossil fuels, representing linear flows of carbon from the ground to the atmosphere, and some may represent circular flows of carbon into and out of the atmosphere. A 2021 National Academies report on accelerating deep decarbonization in the United States estimated that there would be a constant level of remaining CO_2 emissions at about 5 percent of 2005 levels, or about 300 MMT CO_2, in a net-zero emissions scenario for 2050 (NASEM 2021a). The same report also noted that two comprehensive modeling studies of pathways for decarbonization each included >500 MMT of carbon capture and storage (CCS) in the United States per year (DDP n.d.; Larson et al. 2021). A 2022 study of active carbon management summarized several modeling studies' requirements for carbon removal, finding a range of ~500 MMT to ~8 Gt carbon removal required annually worldwide to reach net-zero goals (ITIF 2022). Another method to estimate the possible scale of remaining emissions in a decarbonized future is to consider the "hard to abate" sectors such as aluminum, plastic, steel, and cement, as addressed by the Mission Possible study. That study estimated that the remaining global emissions for those products in a circular carbon scenario would be 5.6 Gt CO_2 per year, 0.8 from aluminum, 0.9 from plastics, 1.9 from steel, and 2.0 from cement (Energy Transitions Commission 2018).

The remaining CO_2 produced in a net-zero system is likely to have multiple potential fates, including CO_2 utilization for durable storage (Track 1; see Figure 1-3-1) or circular processes (Track 2), but also CCS or non-utilization negative emissions processes such as land- or ocean-based enhanced weathering. Remaining fossil emissions can only be used sustainably for durable Track 1 products, while biogenic, direct air capture (DAC), and direct ocean capture (DOC) sources can be used sustainably for either Track 1 or Track 2 products. LCAs of complete CO_2 utilization systems will determine a particular product's carbon footprint, which includes considering the emissions impact of the CO_2 source, as it relates to the product's life cycle from sourcing of materials through manufacture, processing, use, and disposal. Such analyses require materials and processes to be accounted for and attributed, but that does not necessarily mean that different CO_2 sources must be segregated according to their emissions impact. As with other commodities, sources of CO_2 can be mixed, and the origin and fate of any particular molecule of CO_2 need not be specifically known, as long as reliable bookkeeping allows for appropriate attributions to be made and the balance known and accounted for.

For the circular processes generating Track 2 products, where carbon from CO_2 replaces fossil-origin carbon, CO_2 utilization will compete with other potential sustainable carbon feedstocks including biomass and recycled materials. The interplay between these carbon feedstocks is beginning to be explored, particularly with regard to their role in "de-fossilizing" the chemical industry (Bazzanella and Ausfelder 2017; Gabrielli et al. 2020; Lange 2021). Recycling used carbon products can be a first step toward a circular economy, but since it inherently cannot achieve 100 percent efficiency, meeting the demand for chemical products will require additional carbon inputs from biomass or CO_2 (Lange 2021). Among different routes to produce a carbon-neutral product, trade-offs exist between requirements for land, scarce materials and minerals, and water (Gabrielli et al. 2020). Efforts toward developing a circular economy are particularly strong in Europe, as evidenced by the European Commission's Circular Economy Action Plan released in 2020 (EC 2020). A detailed analysis of the European chemical industry found that achieving carbon neutrality to meet the European Commission's larger climate goals would require innovative research and development efforts, including through long-term public–private partnerships to facilitate deployment and de-risking of new technologies (Bazzanella and Ausfelder 2017). Other recommended policies to promote de-fossilizing the European chemical industry included establishing central, open-access databases for (1) LCA studies of low-carbon technologies and (2) carbon sources (biomass, CO_2, and other gases) and existing infrastructure (Bazzanella and Ausfelder 2017).

3.3 POTENTIAL UTILIZATION PRODUCTS AND PROCESSES

Going forward, transformations of CO_2 are likely to expand dramatically beyond the current limited set of chemically or biologically advantageous chemicals, described above, in order to synthesize many net-zero or net-negative carbon-containing products. Table 3-2 and the section that follows summarize classes of materials to consider for future CO_2 utilization, drawing from a comprehensive research status assessment published by the National Academies in 2019. When planning infrastructure for future CO_2 utilization, however, it is useful to consider not only what products can be made from CO_2, but also what products will be a high priority in a future net-zero economy, and what fraction of the product market will be made with carbon from biomass and recycling. Section 3.5 highlights priority technologies for a net-zero future.

TABLE 3-2 Summary of Major Potential CO_2 Utilization Opportunities Including Utilization Process and Infrastructure Aspects

Chemical or Material	Utilization Process	Summary of Infrastructure Aspects by Product Class
Construction material		
Concrete	Mineralization	• Product is a bulky solid commodity. Pourable mixtures have a short lifetime curing time before they must be used, limiting transportation. • Some technologies can accommodate impure CO_2 streams. • May be integrated directly onto a flue gas stream. • Localized production. • Track 1: Can sustainably use fossil-derived CO_2.
Aggregates	Mineralization	• Product is a bulky solid commodity, used locally. • Some technologies can accommodate impure CO_2 streams. • May be integrated directly onto a flue gas stream. • Localized production. • Track 1: Can sustainably use fossil-derived CO_2.
Chemicals and fuels		

C1 Compounds
• Methanol
• Formic acid
• Formaldehyde
• Methane
• Carbon monoxide
C2 Compounds
• Ethanol and ethylene
• Dimethyl ether
• Oxalate and oxalic acid
C2+ Compounds
• C2+ carboxylic acids and carboxylates
• Hydrocarbon fuels
• Protein
• Pigments
	Chemical and biological	• Product is commodity chemical for use in the manufacture of other products, or for direct use, often as a fuel.
		• Chemical production:
		  ◦ Centralized, large-scale production, possibly co-located with production of other chemicals and products.
		  ◦ Often will require significant supplies of either hydrogen (similar order of magnitude as CO_2 supply) and/or electricity.
		  ◦ Often requires high-purity CO_2.
		• Biological production:
		  ◦ Small-scale production.
		  ◦ Requires availability of water supply.
		• Track 2: Must use CO_2 derived from the atmosphere or biosphere for sustainability.

Niche Products
• Diamonds
• Vodka
• Other small-scale, high-value products
	Chemical	• Product is often of high value, but produced at small scale.
		• Chemical production:
		  ◦ Often will require significant supplies of either hydrogen (similar order of magnitude as CO_2 supply) and/or electricity.
		  ◦ Often requires high-purity CO_2, especially for food products.
		• Track 1 (lifetime >100 years): Can sustainably use fossil-derived CO_2.
		• Track 2 (lifetime <100 years): Must use CO_2 derived from the atmosphere or biosphere for sustainability.

continued

TABLE 3-2 Continued

Chemical or Material	Utilization Process	Summary of Infrastructure Aspects by Product Class
Polymers, polymer precursors, and other materials		

Polymers and polymer precursors	Chemical and biological	• Product is used to produce polymers, or products from polymers such as plastics, fibers, and other materials.
		• Chemical production:
		  ○ Centralized, large-scale production, possibly co-located with production of other chemicals and products.
		  ○ Often will require significant supplies of either hydrogen (similar order of magnitude as CO_2 supply) and/or electricity.
		  ○ Often requires high-purity CO_2.
		• Biological production:
		  ○ Smaller-scale production.
		  ○ Requires availability of water supply.
		• Track 1 (lifetime >100 years): Can sustainably use fossil-derived CO_2.
		• Track 2 (lifetime <100 years): Must use CO_2 derived from the atmosphere or biosphere for sustainability.

0-3D elemental carbon and engineered products	Chemical	• Products are solid materials with exceptional properties. They are used for manufacturing various advanced high-value materials for energy generation and storage, environmental protection, catalysis, and structural materials.
		• Chemical production:
		  ○ Tunable production scale, depending on market needs.
		  ○ Often need high reaction temperature, inert gases, and special pressures.
		  ○ Often use specialized production equipment.
		• Track 1 (lifetime >100 years): Can sustainably use fossil-derived CO_2.
		• Track 2 (lifetime <100 years): Must use CO_2 derived from the atmosphere or biosphere for sustainability.

3.3.1 Construction Materials

3.3.1.1 Concrete

Concrete materials offer a particularly attractive opportunity for CO_2 utilization. With an estimated annual global production of 20–40 Gt by 2030 (see Table 3-1) and resulting durable CO_2 use of up to 1.41 Gt per year by 2050, the potential CO_2 utilization could be at a climate-relevant scale. Furthermore, CO_2 use in concrete production results in durable conversion to solid Track 1 materials and thus constitutes a CDR process. CO_2 incorporation into concrete can occur via a variety of mechanisms, including curing via carbonation of the cement or supplemental cementitious material, as well as the use of fillers and aggregates that themselves have been manufactured from minerals or waste materials by carbonation with CO_2 (Hills et al. 2020; Woodall et al. 2019). An example of a mineral carbonation reaction is CO_2 interacting with portlandite, an alkaline mineral, to produce calcite, the mineral carbonate product, and water, $Ca(OH)_2 + CO_2 \rightarrow CaCO_3 + H_2O$. This is a thermodynamically favored and exothermic reaction at ambient pressure and temperature.

The amount of CO_2 incorporated into the material during curing varies substantially between so-called precast concrete products (>1 percent by mass) and ready-mix concrete (<1 percent by mass). Although the methods differ in applications and performance, given suitable production conditions, both can yield concrete materials that exhibit higher compressive strength. In building applications where compressive strength is important, then the same structure could be built with less concrete (Monkman and MacDonald 2017; Ravikumar et al. 2021b). Note that cement curing with CO_2 reduces the pH of the resulting concrete compared to traditional concrete, which could be detrimental if steel reinforcements are used in the concrete, since corrosion of steel is suppressed only at higher pH levels. Solutions to this potential shortcoming could include use of alternative concrete materials that

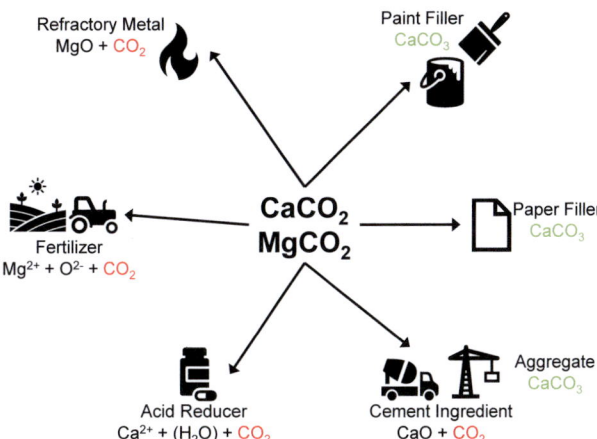

FIGURE 3-2 Utilization opportunities for mineral carbonates.
SOURCE: C.M. Woodall, N. McQueen, H. Pilorgé, and J. Wilcox, 2019, "Utilization of Mineral Carbonation Products: Current State and Potential," *Greenhouse Gases: Science and Technology* 9(6):1096–1113, https://doi.org/10.1002/ghg.1940. CC BY 4.0.

seek to avoid the need for most, if not all, steel reinforcements via increased ductility and self-healing capability (Li 2019) or replacing steel rebar with sustainably sourced and produced carbon fiber bars. Concrete with modified properties often results in conflicts with existing codes, standards, and approved materials lists for procurement, making the introduction of these new concretes more difficult. Additionally, conventional and CO_2-cured concrete absorb CO_2 from the ambient air, so there has been some question of the impact of initial CO_2 curing on the total lifetime absorption of CO_2. Continued curing of concrete during its service life is slow and dependent on many factors, and quantitative understanding of the impact of CO_2-curing concrete during construction on future CO_2 uptake is not well understood (Zhang et al. 2020).

3.3.1.2 Aggregates

As with concrete materials, the carbonation of minerals and selected waste materials offers Track 1–type CO_2 utilization opportunities, that is, providing durable storage of CO_2. An inorganic material with sufficient alkalinity, typically from available Ca^{2+} or Mg^{2+} or their oxides, will react with CO_2 to form durable carbonates that could be considered for use as aggregates, such as in concrete production, as a substitute for gravel, or as fillers in paper or porcelain production, as illustrated in Figure 3-2 (La Plante et al. 2021; Woodall et al. 2019). Materials such as steel slag, cement kiln dust, and fly ash contain a significant amount of alkaline material readily available for direct reaction with CO_2, while some minerals require chemical processing to make the alkaline materials available. CO_2 dissolved in natural bodies of water can also be utilized to form inorganic carbonates. Such processes also have been designed as chemical looping processes with the purpose of capturing CO_2 (Adánez and Abad 2019).

3.3.2 Chemicals and Fuels

Using chemical or biological processes, CO_2 can be converted into various carbon-based chemicals and/or fuels. Present global efforts focus on converting CO_2 into C1 and C2 materials with single or multiple steps. The 2019 National Academies' assessment of CO_2 utilization status and research needs identified major opportunities for chemical and fuel formation via CO_2 utilization. Figures 3-3 and 3-4 show the major products accessible via chemical and biological routes, respectively.

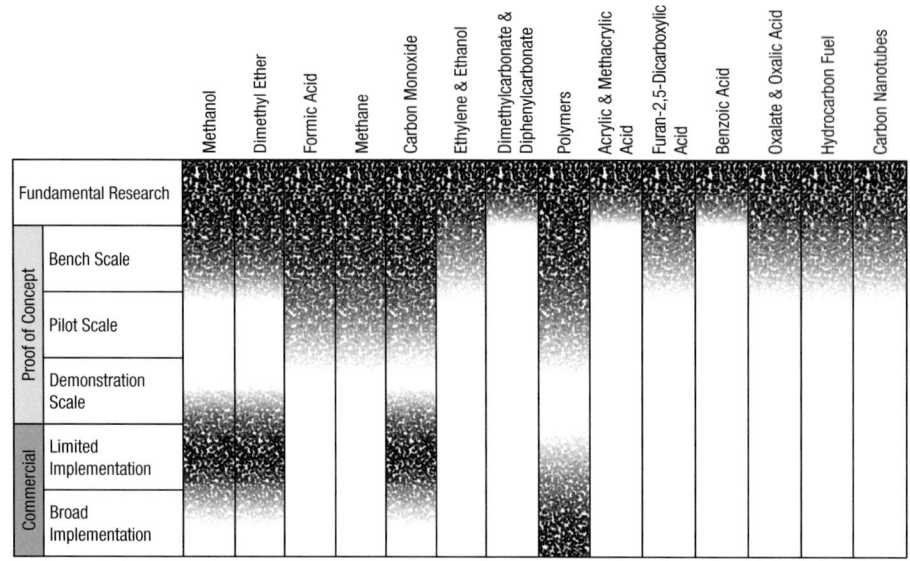

FIGURE 3-3 Chemically accessible products from CO_2, including technical maturity.
SOURCE: National Academies of Sciences, Engineering, and Medicine, 2019, *Gaseous Carbon Waste Streams Utilization: Status and Research Needs*, Washington, DC: The National Academies Press, https://doi.org/10.17226/25232.

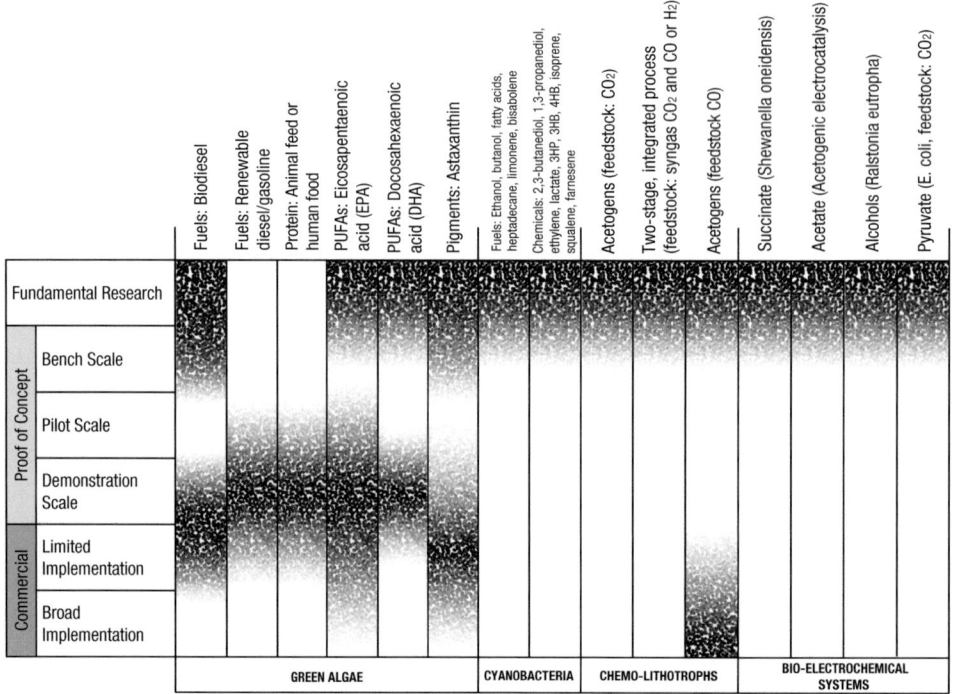

FIGURE 3-4 Biologically accessible products from CO_2 utilization, including technical maturity.
SOURCE: National Academies of Sciences, Engineering, and Medicine, 2019, *Gaseous Carbon Waste Streams Utilization: Status and Research Needs*, Washington, DC: The National Academies Press, https://doi.org/10.17226/25232.

3.3.2.1 C1 Compounds

C1 compounds are important targets for CO_2 utilization because they are major commodity chemicals or intermediates and are more easily chemically synthesized from CO_2 than chemicals that require the formation of carbon-carbon bonds. CO_2 could be converted directly into partially or fully hydrogenated C1 compounds such as formic acid (HCOOH), methanol (CH_3OH), and methane (CH_4), as well as other important C1 products such as carbon monoxide (CO) and urea (($H_2N)_2CO$). Many of these products are used on their own as is or for industrial processes, and some serve as feedstocks for other chemical syntheses.

Urea, an input to fertilizer production and other industrial processes, is currently the largest global chemical consumer of CO_2, and urea synthesis is expected to remain a major consumer. Carbon monoxide is also an important intermediate product of CO_2 utilization, because it may become an important intermediate to sustainable fuel production via Fischer-Tropsch pathways, or for other chemical processes, in a net-zero emissions future. Of the partially and fully hydrogenated products of CO_2, methane and methanol could serve as energy carriers; however, the toxicity of methanol may limit its distributed uses (chemeurope n.d.; IRENA and Methanol Institute 2021). Methanol and formic acid are commodity chemicals and could serve as intermediates for other sustainable chemical and fuel production, including methanol-to-gasoline via, for example, the Mobil zeolite catalysis process.

3.3.2.2 C2+ Compounds

C2+ compounds are important targets for CO_2 utilization because they constitute fuels, commodity chemicals, and chemical feedstocks. Key C2 compound targets are ethylene, ethanol, oxalate, and oxalic acid. Other C2+ compounds that could be synthesized from CO_2 include alcohols, carboxylic acids, fuels, pigments, and proteins. Both chemical and biological systems can produce C2+ chemicals, with biological systems being especially adept at forming multicarbon products. For example, a process using microbes to convert CO_2 to ethanol has been commercialized (Köpke et al. 2011; McCoy 2022). In addition to ethanol, C2+ products accessible via biological processes include various alcohols and hydrocarbons and their mixtures, diols, and carboxylic acids, particularly carboxylic acids that are also metabolic biomolecules. Both chemical and biological processes can produce individual chemicals and mixtures that may be used as fuels, such as C2–C8 hydrocarbons and alcohols. Processes to create synthetic fuels and biofuels from CO_2 can serve the potential markets for sustainable fuels for heavy-duty transportation, particularly aviation, and for storage of renewable energy in chemical bonds (so-called Power-to-X).

3.3.3 Polymers, Polymer Precursors, and Composites

Polymers and polymer precursors are high-demand commodities. The global market for polymers is expected to grow 5.1 percent annually and will reach more than $1 trillion by 2030 (Research and Markets 2020). The major driver of the rapid market growth has been the fast transfer from reusable packaging to one-time utilization of polymer-based plastics. While there are efforts to reduce single-use plastics, some single-use, disposable plastic requirements will remain, especially for medical applications. Other polymer uses include in fibers, rubbers, and fabrics, and for some durable plastic applications, such as materials in vehicles and appliances.

Chemical syntheses of polymers from CO_2 can occur via direct or indirect routes. CO_2 is directly incorporated into aliphatic polycarbonates as carbonate groups via copolymerization with epoxides, or indirectly incorporated via conversion to other monomers, such as methanol, carbon monoxide, ethylene, organic carbonates, or urea (NASEM 2019). CO_2 integration into polyols to produce polyurethanes has been demonstrated and scaled to pilot-size plants (Langanke et al. 2014). Benzene may be another polymer precursor of interest, as it is currently used to synthesize derivatives that are polymerized to form polystyrene and synthetic rubber, among others. Only a part of the carbon in polycarbonates is CO_2-derived when CO_2 is copolymerized with epoxides. Research is exploring renewable feedstocks such as cyclohexadiene oxide, limonene oxide, and α-pinene oxide. CO_2-derived polymers can be modified for more applications by blending with natural and synthetic compounds, integrating with fillers, and crosslinking (e.g., vulcanization).

In addition to chemical means of producing polymers from CO_2 utilization, biological systems can produce polymer precursors and biopolymers. Biologically derived polymer precursors can be used to synthesize polyesters, typically via an esterification reaction using a four-carbon dicarboxylic acid monomer and a diol monomer. Many dicarboxylic acids have been thoroughly investigated as valuable fermentation products that can be made efficiently in various microorganisms. Most of these dicarboxylic acids are found in the citric acid cycle and include succinate, fumarate, and malate. Other carboxylic acids of note include 2,5-furan dicarboxylic acids such as 2,5-dioxopentanoate. Research into the microbial production of the diol monomers is still in its infancy, including of 2,3-butanediol and 1,3-propanediol, but the promise of making industrially viable bioplastics has led to increased research investigating the production of various diols. Other biological polymer precursors include ethylene, for making polyethylene (NASEM 2019). Cyanobacteria can also produce biodegradable polymers, especially polyhydroxy alkanoates, and work is ongoing to produce this material from CO_2, rather than sugars (Troschl et al. 2017).

3.3.4 Elemental Carbon Materials

Conventional and advanced carbon materials with 0-3D structures, for example, carbon quantum dots, carbon nanotubes, graphene, graphite, amorphous carbon, carbon fiber, carbon black, and carbon-carbon composites, offer opportunities for advanced properties and new uses. The carbon materials are stable, although their stabilities are affected by the conditions in which they are utilized. The markets of carbon quantum dots, carbon nanotubes, graphene, graphite, amorphous carbon, carbon fiber, carbon black, and carbon-carbon composites in 2020 were $0.652B (Mordor Intelligence 2021), $0.877B (Emergen Research 2022), $0.286B (Fortune Business Insights 2022), $13.6B (Fortune Business Insights 2021), $0.226B (Research Reports World 2021a), $16.0B (Global Market Insights 2021), $2.523B (Research Reports World 2021b), respectively. Markets for carbon materials are expected to rapidly increase. A recent study of CO_2 utilization potential that examined carbon black estimated a market value of $14B–$66B in 2050 and a potential for CO_2 use of between 40 and 200 MMT (Sick et al. 2022b). Outside of carbon black production, technologies for the conversion of CO_2 to other carbon materials are less mature, and market projections are difficult. Carbon materials are rich in structure, texture, and properties. Zero-dimensional, or 0-D, carbon materials include graphene quantum dots, carbon quantum dots, nano-diamond, fullerenes (e.g., C60), carbon black, and carbon-coated metal nanoparticles. Representative 1D, 2D, and 3D carbon materials include carbon nanotubes and carbon nanofibers; graphene; and carbon foam and expanded graphite, respectively. Different dimensional materials have correspondingly different applications based on their structure and properties. For example, due to their tiny size, quantum confinement effects, desirable properties, and biocompatibility, 0D carbon materials have great potential in sensing applications (Uygun and Uygun 2020; Wang et al. 2020). 1D carbon nanotubes and 2D graphene are being explored for a plethora of advanced technologies. Three-dimensional hierarchical carbons are inexpensive, lightweight, have a high specific surface area and well-ordered channels, and have outstanding electronic and ionic conductivity, and thus are promising materials for energy conversion and storage for high-strength material applications. Carbon black applications range from filler materials, for example, for tires and toners, to additives for concrete. Using carbon black as a filler in durable materials offers opportunities for growing Track 1 type utilization of CO_2 with durable carbon storage. Finally, carbon-carbon composites, made from carbon fiber and graphite, are lightweight and very strong, enabling their use as structural materials in vehicles and buildings.

3.3.5 Niche Products: Diamonds, Perfumes, and Liquor

Consumer-facing products made with CO_2 might not have a large CO_2 utilization potential by volume but can serve as signaling products to increase consumer familiarity and comfort with CO_2-derived products (Arning et al. 2020; Lutzke and Árvai 2021). Examples of such products include diamonds, high-purity ethanol for the production of perfumes and vodka, personal care products, and food (Aether n.d.; Air Protein 2022; Gallon 2022; Kiverdi 2019; Köpke et al. 2011; Pace and Sheehan 2021). In some cases, these products also may provide a greater profit margin than the respective commodity chemicals or intermediates and may allow producers to successfully commercialize other higher-volume processes in tandem to bring down overall costs to compete in a larger market.

3.4 EMERGING, PILOT, AND COMMERCIAL FACILITIES UTILIZING CO_2

In 2016, less than 200 active entities were identified as working on CO_2 utilization (Global CO_2 Initiative 2018) and respective technology readiness levels were mostly below 5. Since then, more developers have entered the field, but the numbers are only now beginning to grow more rapidly. AirMiners and the Circular Carbon Network keep listings of carbon capture, utilization, and storage (CCUS) developers, which include 403 companies (as of May 13, 2022) that work on capture technologies or services, or utilization (AirMiners 2022; Circular Carbon Network 2022). A visual database that includes additional information, such as publications and policies, is also available (Global CO_2 Initiative 2022b). Figure 3-5 shows locations of startup CO_2 utilization companies in the United States.

3.5 PRIORITY NEEDS FOR CO_2-DERIVED PRODUCTS THAT COULD CONTRIBUTE TO A NET-ZERO CARBON FUTURE

3.5.1 CO_2-Derived Product Priorities

Products from CO_2 utilization that will be of highest importance and significance in a net-zero carbon future (IPCC 2022) include Track 1 materials that offer durable CO_2 storage and Track 2 materials for which the climate benefit comes from replacing fossil carbon with CO_2-based carbon in a circular fashion for carbon-containing products that continue to be necessary and desirable. These products are summarized in Table 3-2, above. Even-

FIGURE 3-5 Locations of startup CO_2 utilization companies in the United States listed in the Circular Carbon Database. Note that no startup CO_2 utilization companies are listed for Alaska or Hawaii.
SOURCE: Global CO_2 Initiative, 2022, "CCU Activity Hub," University of Michigan, https://www.globalco2initiative.org/evaluation/carbon-capture-activity-hub.

tually, products in both tracks could conceivably displace the incumbent products completely. However, this outcome will depend on many factors related not only to manufacturability, cost, and other externalities, but also to the development of alternative processes and products that may replace incumbent fossil-carbon products. Examples of such alternative products include electricity and clean hydrogen as energy carriers for mobility applications.

3.5.2 Requirements for Other Inputs and Feedstocks

Success for CO_2 utilization depends not only on access to CO_2, but also on an array of other inputs and feedstocks, all of which must contribute to a net-zero or net-negative system. In particular, CO_2 utilization systems require scale-up of carbon-emission-free energy for both utilization processes and other parts of the CO_2 utilization system, including capturing and processing (separating and purifying) CO_2, generating other inputs, and transporting reactants and products. Additionally, some processes may require clean hydrogen, water, and land. Large-scale deployment of CO_2 conversion technology could be problematic if the technology relies on catalysts that require difficult-to-source and -secure access to raw materials (including some metals).

3.5.2.1 Electricity and Fuel Requirements for Carbon Capture

Figure 3-6 shows the electric and fuel energies, which, summed together, are required to capture CO_2 from various sources. For point sources, the amount of electric energy and/or fuel required for carbon dioxide capture and storage (CCS) is taken from the 2019 National Petroleum Council Carbon Capture study. For DAC, the energy requirements for liquid and solid capture come from IEA (2022).

Capital costs for carbon capture depend on CO_2 concentration and pressure of the waste gas stream and vary by more than 10-fold for the industrial processes shown. For all technologies other than DAC, the energy required

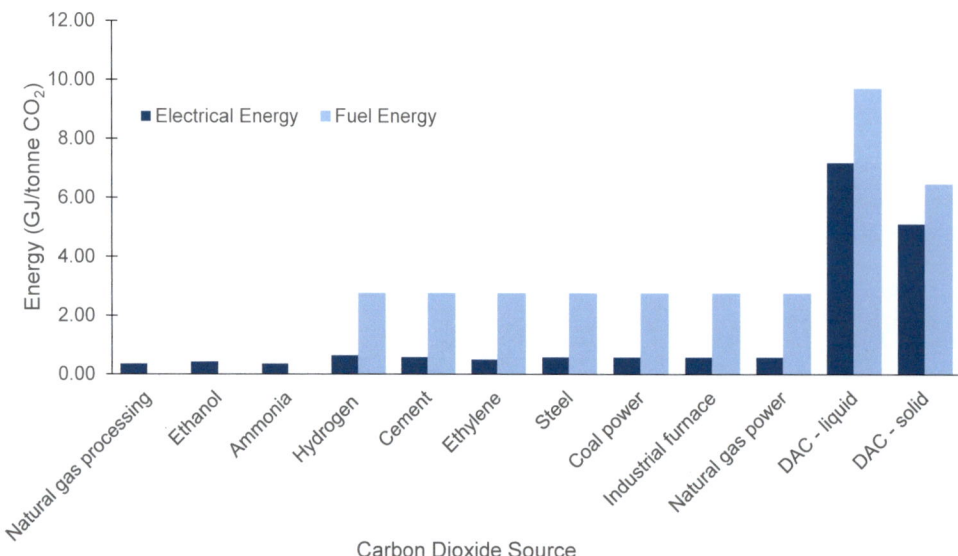

FIGURE 3-6 Electrical and fuel energy requirements for CO_2 capture from different sources.
SOURCES: Committee generated based on data from National Petroleum Council, 2019, "Meeting the Dual Challenge: A Roadmap to At-Scale Deployment of Carbon Capture, Use, and Storage," https://dualchallenge.npc.org; and International Energy Agency, 2022, *Direct Air Capture: A Key Technology for Net Zero*, Paris: IEA, https://www.iea.org/reports/direct-air-capture-2022. All rights reserved; as modified by the National Academies of Sciences, Engineering, and Medicine.

for CO_2 capture is less than 20 percent of the energy released via fossil-fuel combustion that generates the CO_2 (e.g., 18 GJ/ton of CO_2 released from methane combustion). Because of very low (~420 ppm) CO_2 concentrations in air, DAC requires a greater amount of energy than post-combustion carbon capture (IEA 2022). Current DAC research therefore is examining use of waste heat, or pressure or humidity swings, to improve the return on energy investment. Figure 3-6 shows that selecting a CO_2 source is important in assessing the electric power and fuel requirements for capture. CO_2 also must be captured from the fuel used in process heating and electricity generation. In the future, as electrical grids decarbonize, low- or zero-carbon electricity may replace the fuel portion of energy for capture (Müller et al. 2020).

3.5.2.2 Hydrogen Requirements for CO_2 Utilization Products

Once the CO_2 is captured, many different products can be made. Figure 3-7 depicts the inputs of hydrogen required for various classes of products formed based on reaction stoichiometries. Less hydrogen is required for materials that incorporate oxygen into the final product.

3.5.2.2.1 Electricity and Feedstock Requirements for Hydrogen Production

Figure 3-8 shows a tabulation of the amount of water, electricity, and natural gas to make clean hydrogen via water electrolysis, steam methane reforming (SMR), autothermal reforming (ATR), partial oxidation (POx) of methane, and methane pyrolysis (MP). Multiplying these inputs per unit of hydrogen by the hydrogen consumption per ton of product (see Figure 3-7) yields the inputs required per ton of final product. Note that hydrogen can be manufactured on-site or remotely and supplied by pipeline, truck, or rail. Chapter 4 discusses the infrastructure needs to deliver hydrogen and electricity to the utilization site.

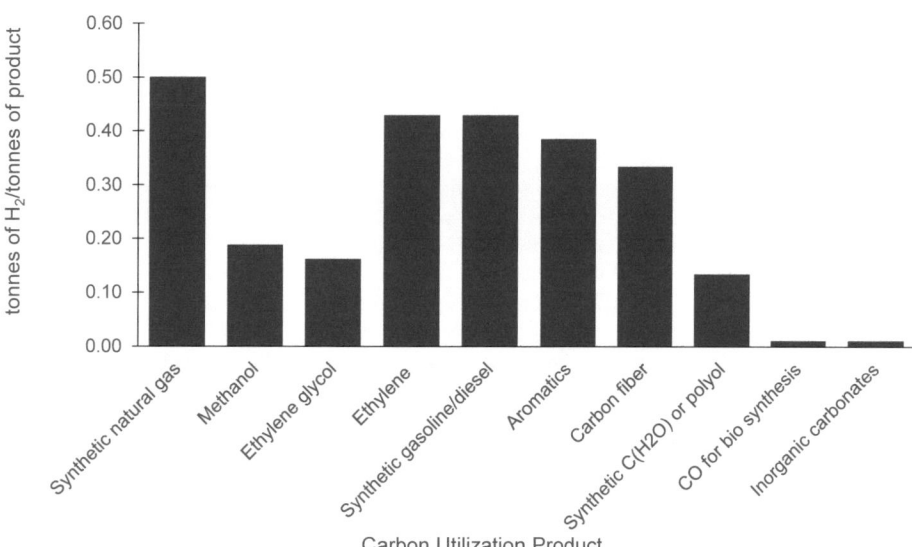

FIGURE 3-7 Stoichiometric hydrogen consumption for CO_2-derived products. Process routes were taken from literature (Agora Verkehrswende et al. 2018; Bazzanella and Ausfelder 2017) or via reaction stoichiometries. Electricity required for hydrogen generation is not included. Assumes high-yield industrial processes with the limit of 100 percent yield assumed in the calculations. The estimated hydrogen required for carbon fiber is for hydrogen reduction of CO_2; electrolytic or pyrolytic dissociation to form carbon fiber and O_2 would require no hydrogen. Biosynthesis and inorganic carbonates require no hydrogen. Synthetic $C(H_2O)$ represents a hydrocarbon chain with a ratio of two hydrogens to every carbon and oxygen.

The endothermic SMR process involves the following chemical reaction:

SMR: $CH_4 + H_2O \rightarrow CO + 3H_2$.

The exothermic partial oxidation (POx) reaction is

POx: $CH_4 + \tfrac{1}{2} O_2 \rightarrow 2H_2 + CO$.

Autothermal reforming seeks to balance exothermic and endothermic reactions to be energy neutral (Luque and Speight 2015):

ATR: $CH_4 + \tfrac{1}{4} O_2 + \tfrac{1}{2} H_2O \rightarrow CO + \tfrac{5}{2} H_2$.

CO produced in the above processes can be reacted with water to form more H_2 via the mildly exothermic water-gas shift (WGS) reaction:

WGS: $CO + H_2O \rightleftharpoons CO_2 + H_2$.

Though this section is about production of hydrogen as an input to CO_2 utilization, it should also be noted that the WGS reaction is a reversible equilibrium and can be used to convert captured CO_2 to CO via the reverse, or "backward," reaction. CO in mixtures with H_2 constitutes "synthesis gas," from which the entire fuels and petrochemical economy can be reproduced via Fischer-Tropsch synthesis (FTS) or methanol synthesis (plus subsequent conversion reactions). FTS and methanol synthesis are fully developed commercial processes that have been practiced for more than 100 years, at scales equivalent to full commercial petroleum refineries (e.g., gas-to-liquids production in Qatar). While catalysts are known and demonstration facilities have been operated, the reverse WGS reaction is the only step that is not fully optimized and practiced at commercial scale to enable CO_2 utilization, but technology readiness is already at a high level. Existing technology thus is proven and demonstrated for use of CO_2 as a full replacement for the fossil-based fuels-and-chemicals economy, if desired.

To avoid formation of gaseous CO_2 product, endothermic MP can be conducted to form solid carbon via

MP: $CH_4 \rightarrow C + 2H_2$.

The process has been practiced for nearly a century to form carbon black or sometimes hydrogen. Current research and development (R&D) seeks to efficiently and continuously co-produce both, including formation of structured carbons for building materials and generation of hydrogen as an energy carrier for the clean energy input needed to drive the endothermic reaction. A portion (50 percent) of the hydrogen generated can be used for this purpose. Technology varies widely from thermal to thermo-catalytic, to plasmas.

The above stoichiometric reactions do not describe the full inputs required to generate hydrogen. Required water inputs can increase up to 10-fold above the stoichiometric need due to water purification (e.g., for electrolysis), treatment, and process cooling steps (Blanco 2021; Lampert et al. 2015, 2016), as well as the need to inject additional water (steam) to prevent coking of catalysts and equipment for natural gas–based synthesis. A similar problem holds for energy and hydrogen inputs above the thermodynamic or stoichiometric minimum. SMR requires energy inputs to run process equipment and heating steps for carbon capture, in addition to driving the endothermic reaction itself. ATR seeks to combust a portion of the natural gas feed to provide the endothermic reaction heat, but also requires electricity and heat (for capture solvent/sorbent regeneration) to run carbon capture, as well as water feed and final product purification steps. POx combusts additional natural gas feed to produce excess heat, which is used to generate electricity to power all process steps, such that all CO_2 is generated in one stream for capture, reducing capital costs (Liu 2021). If individual input materials and processes are net-emitting, then they can be paired with a separate, verifiable negative emissions process (e.g., DAC plus storage), which will allow the entire system to be net-zero or net-negative emissions.

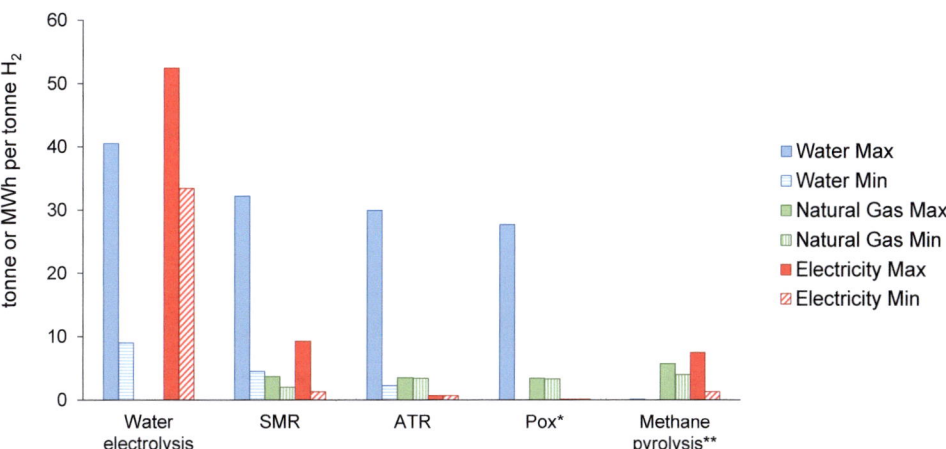

FIGURE 3-8 Water (tonnes, blue), natural gas (tonnes, green), and electrical energy (MWh, red) required to produce hydrogen via water electrolysis, steam methane reforming (SMR), autothermal reforming (ATR), partial oxidation of methane (POx), and methane pyrolysis. Solid bars show maximum values and patterned bars show minimum values, in tonnes or MWh per tonne H_2. * = generates excess electricity; ** = uses H_2 for endothermic energy. Maximum electricity and/or hydrogen and natural gas feedstock and process energy inputs are based on industry reports (Liu 2021), including those compiled by IEA (IEAGHG 2017). Future electrolyzer system efficiencies were estimated at 75 percent as targeted for 2030+ by Dechema (Bazzanella and Ausfelder 2017) and U.S. DOE (DOE-HFTO n.d.).
SOURCES: Committee generated using data from Bazzanella and Ausfelder (2017); Beswick et al. (2021); Blanco (2021); Collodi et al. (2017); Hopcroft and Papadamou (2021); IEAGHG (2017); Lamb et al. (2020); Lampert et al. (2015, 2016); Liu (2021); Marchese et al. (2020); Oni et al. (2022); Tenhumberg and Büker (2020); and Timmerberg et al. (2020).

3.5.2.3 Enabling Inputs Required for Mineral Products

Steel slag, cement kiln dust, fly ash, or some mined minerals (serpentines) can react exothermically with CO_2 to create valuable products (cements, aggregates). Crushing, grinding, and water exposure pretreatment of native minerals makes suitable cations available for reaction with CO_2 and requires relatively little energy. The reaction between CO_2, water, and the pretreated minerals is exothermic and also requires little energy input, so the total energy consumption needed for CO_2 mineralization is low (Dipple et al. 2021; Woodall et al. 2019). As discussed further in Section 4.4, concrete and aggregate production and use are highly distributed due to the difficulty in transporting both inputs and produced materials, which are large volume and heavy, and the inherent distribution of the final products, which are used in a wide variety of buildings and infrastructure, including roads and other major public works projects.

The primary enabling infrastructure requirement for carbon mineralization processes is the heavy-duty power for mining and comminution of minerals to a particle size that can react with CO_2 at acceptable rates. Hydrogen could be a zero-carbon fuel option for mining trucks and equipment, as battery-electric vehicles may lack the power density needed for the very heavy-duty cycles of equipment. Hydrogen can be supplied directly by pipeline to a refueling terminal, or can be generated on-site by supplying water and clean electricity from the grid. Running handling equipment and operations will require some additional electricity. For solids handling, these power requirements are generally higher per ton of product than from a facility producing a gas or liquid product.

3.5.3 Consumer Sentiment

Public awareness of the origin, production methods, and even the component ingredients in many everyday products is low. Acceptance or rejection of products and services therefore might be guided by perceptions and misconceptions. It is thus critical to include research on societal aspects in any planning for technology development and deployment (Buck 2016). Risk-benefit assessments point to CCU acceptance as being influenced by perceptions of CCS (Arning et al. 2020, 2021; Engelmann et al. 2020). Nature-based solutions for carbon removal might find preferential approval over technology (Wolske et al. 2019). A representative survey of adult U.S. citizens showed that a two-thirds majority would be inclined or comfortable with using CO_2-made products. Subtle differences with respect to product type, gender, and political preference of those surveyed had a mild impact on the responses (Lutzke and Arvai 2021).

3.5.4 Future Market Volumes and Current Price Points

Estimates for market penetration (rates) and price points for CO_2-derived products depend on supporting policy, competition from bio-based carbon sources, and alternative solutions, for example, electricity or hydrogen as an energy carrier instead of hydrocarbons. CO_2 utilization will compete with other sources of carbon (biomass, recycling) as well as other technologies to replace the current use of carbon-based products, for example, electricity or hydrogen for mobility applications. Table 3-1 describes various market projections for products of CO_2 utilization.

Projections for market volumes and CO_2 utilization potential vary widely between published studies, sensitively reacting to the assumptions made regarding technologies, cost, and other competing factors. A recent study projects CO_2 utilization for production of aggregates and precast concrete to grow substantially by 2050, reaching between 1 and nearly 11 gigatonnes per year (see Figures 3-9 and 3-10). While market shares for aggregates are projected to only reach 14–18 percent by 2050, CO_2-cured precast concrete could capture as much as 75–80 percent of the market in 2050. Figure 3-11 shows projections of market size through 2040 for various CO_2 utilization products, including building materials, fuels, polymers, chemicals, protein, and carbon additives.

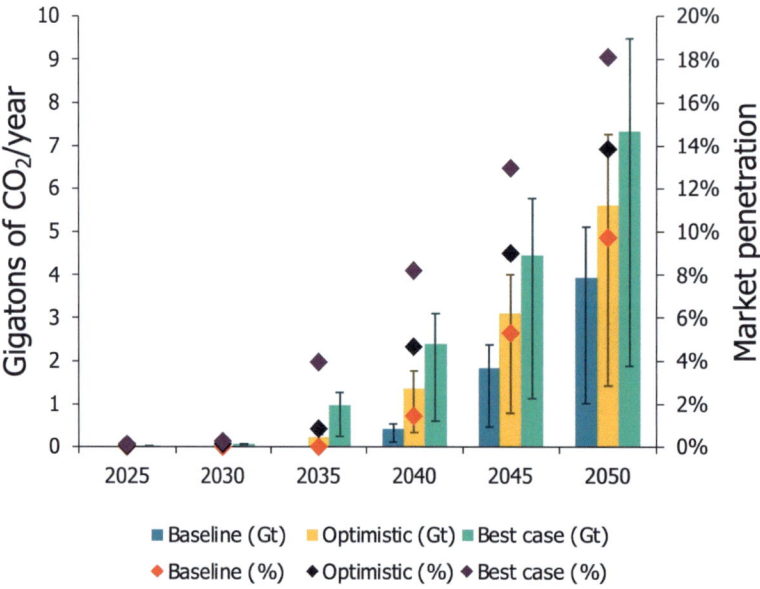

FIGURE 3-9 Projected market penetration and CO_2 utilization potential for aggregates.
SOURCE: V. Sick, G. Stokes, and F.C. Mason, 2022, "CO_2 Utilization and Market Size Projection for CO_2-Treated Construction Materials," *Frontiers in Climate* 4(May):878756, https://doi.org/10.3389/fclim.2022.878756. CC BY 4.0.

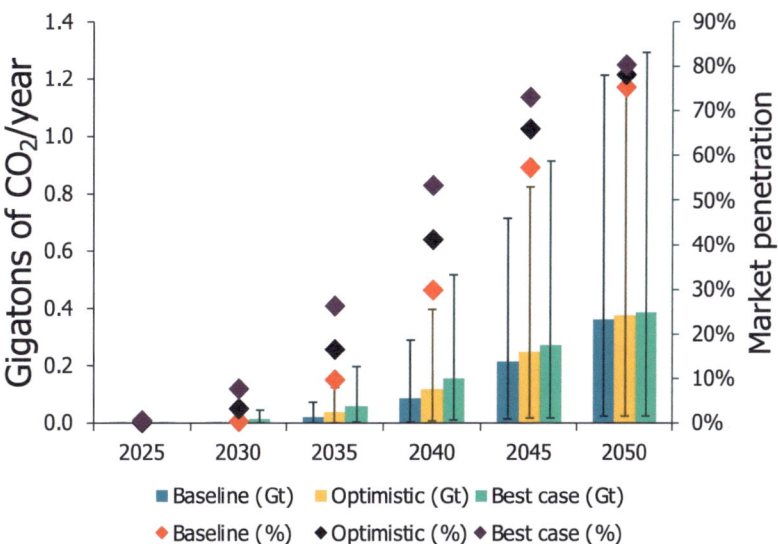

FIGURE 3-10 Projected market penetration and CO_2 utilization potential for precast concrete.
SOURCE: V. Sick, G. Stokes, and F.C. Mason, 2022, "CO_2 Utilization and Market Size Projection for CO_2-Treated Construction Materials," *Frontiers in Climate* 4(May):878756, https://doi.org/10.3389/fclim.2022.878756. CC BY 4.0.

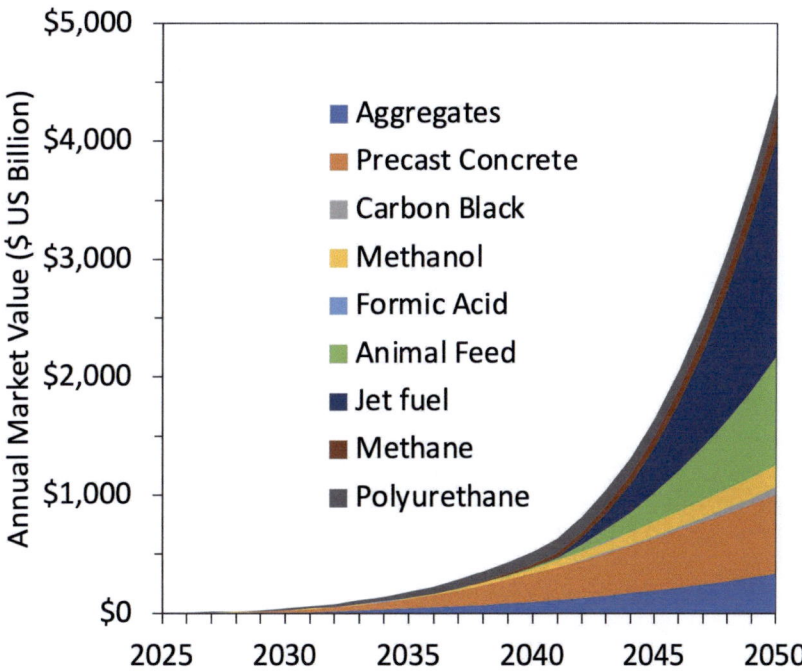

FIGURE 3-11 Market size projections for several major CO_2-derived products through 2050.
SOURCE: Committee generated based on data from V. Sick, G. Stokes, F. Mason, et al., 2022, *Implementing CO_2 Capture and Utilization at Scale and Speed: The Path to Achieving Its Potential*, Technical Report, Global CO_2 Initiative, University of Michigan, https://dx.doi.org/10.7302/5825. Reproduced with permission, Global CO_2 Initiative. CC BY 4.0.

3.6 NEAR-TERM OPPORTUNITIES, SYNERGIES, AND NEEDS

Significant opportunities exist to use CO_2 successfully at large scale to make products. While all of these products can have commercial value, the climate benefit from their production depends on their lifetime, as noted in Section 3.1. Categorizing products as Track 1 or Track 2 is a convenient way to identify what the climate benefits of a given product will be. Responsible launch of the CO_2 utilization industry requires that life cycle assessments in combination with societal impact and techno-economic assessments accompany and guide decision making from early stages of research to commercial deployments.

A CO_2 utilization industry as a whole does not yet exist, but components of it are emerging. Technology readiness for several product categories is at a level where commercial deployments are imminent. Supply and value chains are not established, and work is needed to ensure that infrastructure investments for CO_2 capture and transport strategically and synergistically enable CO_2 capture and utilization. Specifically, the government support enacted in the Infrastructure Investment and Jobs Act of 2021, calling for hubs for DAC and hydrogen, could be leveraged for synergies with CO_2 utilization. Developing such synergies may include, for instance, co-locating at least one DAC and hydrogen hub pair, or locating hubs in regions with an industrial focus to facilitate the adoption of CO_2 utilization and spur private investment. DOE's Notice of Intent for the hydrogen hubs calls for such integration with industrial uses of hydrogen, which could include CO_2 utilization: "End-use diversity—at least one hub shall demonstrate the end-use of clean hydrogen in the electric power generation sector, one in the industrial sector, one in the residential and commercial heating sector, and one in the transportation sector" (DOE-OCED 2022). Such co-located hubs and other concentrations of CO_2 utilization could create nuclei where private investment may find a need for pipelines and other shared infrastructure. Centralized production and use of various fossil fuels and derived chemicals is the norm in the existing petrochemical industry and could be best for CO_2 and hydrogen production and use as well.

The largest amounts of CO_2 could be used to produce construction materials, polymers, fuels, and commodity chemicals. Track 1 CO_2 utilization may serve a need for consuming and storing large volumes of CO_2 that would otherwise be emitted to the atmosphere, such as in construction materials and some durable products. Track 2 CO_2 utilization will be a part of ensuring access to necessary chemicals in a circular carbon economy, including for large volumes of fuels, polymers, and commodity chemicals. The following sections detail opportunities, synergies, and needs for the important, large-volume products in Track 1 and Track 2.

3.6.1 Construction Materials

Opportunities for large-scale CO_2 utilization potential exist for the production of Track 1 construction materials that constitute a means of durable CO_2 storage, equivalent to CCS, yet with the potential for higher societal value and reduced cost.

CO_2 utilization that produces Track 1 construction materials can create synergies to address multiple pressing societal needs given the urgent need to rebuild a substantial portion of U.S. built environment infrastructure, addressing the acute housing shortage and the significant carbon footprint of the built environment. It is instructive to note how slowly durable products penetrate the market due to 10+-year use phases for legacy technology. Rapid launch of CO_2-based construction materials is therefore essential, since infrastructure investments are made for even longer-term use, and CO_2 removal opportunities would be lost with delayed deployments. Mineralization reactions such as by-product use of steel slag or fly ash to reduce energy requirements for concrete (IEA 2019; Mangan 2010) or mining of serpentine minerals for production of aggregates represent some of the most likely options for viable CO_2 utilization with very large potential for CO_2 removal.

To establish CO_2 utilization for concrete and aggregate production, distributed CO_2 sources will be needed, given the fragmented and very localized production requirements for this sector. Co-location with CO_2 point sources needs to be explored as much as possible to reduce transportation requirements. Properties of the new construction materials may differ from incumbents, triggering the need for testing and validation of the new materials, creation of new environmental product declarations, and adaptation of building codes and standards. The latter is nontrivial, given the conservative nature of building codes.

3.6.2 Synthetic Fuels

CO_2 utilization to produce synthetic fuels would provide energy-dense hydrocarbons derived from renewable, rather than fossil, carbon to power existing and future combustion processes. Synthetic fuels are a Track 2 chemical, and therefore their production must use DAC, DOC, or biogenic CO_2 sources because circular carbon flows are required for sustainability. Synthetic fuels may be particularly important for applications where hydrogen- or electric-powered vehicles will be challenging to produce or operate, such as in long-distance, heavy-freight applications in aviation and shipping. Synthetic fuels may also offer a daily, monthly, or seasonal energy storage opportunity. Additionally, local synthetic hydrocarbon fuel production may provide safe and secure access to fuels during overseas deployments, a potentially important application as continued use of existing defense-related vehicle fleets is a significant concern for national security. Such production methods could be cost-competitive relative to logistics costs of supplying today's fuels (Poland 2021). Finally, developing CO_2 utilization for synthetic fuel production affords an opportunity not only to use existing infrastructure but also existing workers, many of whose skills will be translatable from the current fossil-fuel system to a synthetic fuel system.

Despite the opportunities and synergies in using aspects of the current energy system, synthetic fuels will be challenged to compete with alternative methods of de-fossilizing or de-carbonizing combustion applications in transportation, industry, and buildings. Synthetic fuels are expensive to produce, because of both the high capital costs of equipment needed for electrolysis and fuel synthesis and the high electricity consumption of the processes (Searle et al. 2021). Other potential sources of sustainable drop-in, energy-dense fuels include biomass- or waste-derived hydrocarbons. All drop-in fuels will compete with hydrogen and/or electricity to power equipment and processes, especially because combustion of hydrocarbon fuels is less efficient for powering transportation or other processes than use of hydrogen for combustion or in fuel cells, or direct electrification. A 2021 study of light-duty vehicle fuel economy examined the emissions and energy-use implications of various alternative fuels compared to conventional internal combustion engine vehicles powered with standard gasoline with 10 percent ethanol. Using the GREET model to estimate energy use of alternative-fueled vehicles, the study found that electric cars use 45 percent of the energy of conventional vehicles, fuel-cell vehicles powered with electrolytic hydrogen generated from renewable energy use 78 percent of the energy of conventional vehicles, and conventional vehicles powered by synthetic fuels synthesized from CO_2 and electrolytic hydrogen using renewable energy use 297 percent of the energy of conventional vehicles (NASEM 2021b).

In addition to the challenges associated with higher energy use, combustion of hydrocarbon fuels is a source of particulate matter, nitrogen oxides, and ozone (leading to smog in the presence of hydrocarbons), which adversely impact human health with disproportionate impact on disadvantaged communities, unlike zero-emission electrical power or hydrogen-powered fuel cells.[2] Because synthetic fuels may be designed to have lower criteria emissions than current fuels such as gasoline and diesel, they typically have fewer contaminants than fossil fuels and can be made as custom blends of chemicals. Aviation fuel or perhaps marine propulsion are possibly the best uses for synthetic hydrocarbon fuels, since local air quality impacts would be less severe and electrification is difficult due to the low energy and power densities afforded by batteries. Particularly for aviation, the ability to use existing aircraft and distribution infrastructure together with a decrease in manufacturing costs for synthetic aviation fuel may enable CO_2-derived synthetic fuels to be nearly cost-competitive with existing hydrocarbon fuels and other alternative hydrogen options by 2050, if there is a price on CO_2 emissions (Third Derivative n.d.).

3.6.3 Commodity Chemicals

CO_2 utilization can be one way to de-fossilize production and use of carbon-based commodity chemicals. Unlike the fuels described above, there are few options for replacing carbon-based commodity chemicals with carbon-free alternatives.

Commodity chemicals from CO_2, such as synthetic fuels, have synergy with the existing infrastructure for distribution and use of the final chemical or material product. The chemical industry is designed to make and use

---

[2] Hydrogen, when combusted rather than used in a fuel cell, can lead to criteria pollutant NOx, if not done judiciously.

several "building block" chemicals to synthesize a large number and volume of chemicals, and so if CO_2 utilization could be used to produce these central intermediates, then much of the same infrastructure and capital could be used for sustainable carbon-based chemical production. Similar to the fuels discussion above, the existing workforce will remain effective for the new circular carbon economy.

CO_2 utilization will compete with other sources of sustainable carbon, including biomass and waste-based resources. For hydrocarbon chemicals production, the same molecules can sometimes be made from bio-based feedstocks with lower specific energy needs, but higher land use (Haider 2022). Whether or not CO_2-derived hydrocarbon chemicals will be competitive with bio-based chemicals will depend on reduction in the cost of DAC and sufficient availability of low-cost, zero-carbon-emission energy. Hydrocarbon chemical products have sequestration lifetimes of less than 100 years (IEA 2019); thus, their chemical synthesis must not involve current fossil emission point sources.

3.6.4 Polymers

Polymers and polymer precursors are a major class of chemicals and materials, used for plastics, rubbers, fibers, and foams. CO_2 utilization has significant potential to displace existing polymer syntheses from fossil fuels and for CO_2 to be a feedstock for other, different polymers optimized for CO_2 utilization. Currently, there is some commercial-scale synthesis of polycarbonates from CO_2, though copolymerization methods and use of fossil-fuel–derived feedstocks result in only a portion of the carbon in the product being derived from CO_2. Biological and chemical systems can produce polymer precursors and polymers.

CO_2 utilization in the production of polymers can fill a need for materials in a circular economy, addressing huge existing markets and offering opportunities to improve polymer syntheses in some cases. However, challenges exist for converting CO_2 to polymers. The reaction conditions for producing polymer precursors, including ethylene and benzene, differ from those of conventional ethylene and benzene production technologies. Thus, the existing ethylene and benzene manufacturing infrastructures need to be modified. Also, the contaminants (trace gases, especially NOx/SOx, and particulates) in the captured CO_2 could lower the activities of the chemical catalysts used in CO_2-to-polymer synthesis. Contaminants in CO_2 feedstock streams are less of a problem for biological CO_2-to-polymer synthesis (Kondaveeti et al. 2020).

Production of polymers from CO_2 in a cost-effective and environmentally friendly manner will require synergistic efforts. First, CO_2 capture and purification will need to be available to provide high-purity CO_2, which not only will improve CO_2 utilization efficiency but also will increase the purity and productivity of polymers and thus the marketability of CO_2-derived polymers. Co-location of CO_2 and hydrogen sources could lower chemical polymer production costs. The choices for specific chemical and biological pathways need to be considered together with the capability of the existing polymer synthesis infrastructures. Polymers are a high-volume material class, accessible from CO_2, recycled material, and biomass feedstocks. CO_2 utilization will be favorable for certain polymer syntheses, including those of polycarbonates and polyurethanes, and will provide an opportunity to produce polymers that replace those lost in recycling processes, which are not 100 percent efficient. Biodegradability of polymers can be an asset for products that are frequently discarded and unlikely to be recycled, and CO_2 utilization can produce those polymers in a circular carbon economy. Some polymers are durable and can represent Track 1–type utilization if products are sufficiently long-lived (>100 years), such as some building materials.

3.6.5 Conclusion

Overall, it is clear that CO_2 utilization is not a silver bullet either to counter the effects of climate change or to meet the needs for carbon-based products. However, it is one of many tools that will be used to accomplish these goals, and in some cases, it is likely to be favorable over alternatives. For example, CO_2 utilization may offer opportunities for CDR at lower cost for Track 1 chemical products compared to CCS, access to Track 2 carbon products with lower land-use requirements than biomass-derived ones, and, for both product tracks, a new industry that may address environmental justice and other negative impacts related to the production of incumbent chemicals and materials. Forecasts for the magnitude of the emerging CCU industry necessarily cover a wide range, but even at lower estimates, in 2050, nearly 2 Gt CO_2 could be utilized in products, generating $1.1 trillion in revenues. If conditions favor CO_2 utilization over other alternatives, a high estimate of 27 Gt CO_2 utilized and $4.4 trillion revenues has been projected. Policy guidance will be essential to ensure an expedited and just launch of the CCU industry.

3.7 FINDINGS AND RECOMMENDATIONS ON POTENTIAL USES OF CO_2 IN COMMERCIAL PRODUCTS

FINDING 3.1 Product Lifetime. Some CO_2-derived products, such as CO_2-cured concrete, carbonated aggregates, carbon fibers used in structural materials, and long-lived polymers, have lifetimes that will lead to durable CO_2 sequestration if handled appropriately. In contrast, short-lived products, for example, fuels, fertilizers, and many other chemicals, will decompose quickly and return the CO_2 to the atmosphere, presenting the opportunity to participate in a circular carbon economy.

FINDING 3.2 Sustainability of Products Based on Carbon Source. Long- and short-lived CO_2-derived products have fundamentally different relationships to carbon sources, resulting in different climate impacts. Long-lived products may be net-negative compatible if non-fossil CO_2 is used, or net-zero compatible if sourced from fossil CO_2. Short-lived products may be net-zero compatible if the CO_2 is sourced from non-fossil sources. If emissions cannot be reduced to be compatible with net-zero or net-negative pathways, such as use of fossil CO_2 for short-lived products, then the resulting product may be considered unsustainable relative to alternative options such as production from biomass or recycled material feedstocks, carbon sequestration, or use of non-carbon-based technologies, such as replacing hydrocarbon fuels with hydrogen or electricity. The long-term sustainability of different CO_2 utilization processes is an important consideration for infrastructure planning and policy.

FINDING 3.3 Carbon Accounting. Carbon accounting across the value chain will be important to assessing sustainability and climate impact of carbon management technologies, including CO_2 utilization.

> **RECOMMENDATION 3.1.** The U.S. Department of Energy should fund research to quantify the dynamic impact of CO_2-derived products, for example, their specific lifetime, on the CO_2 balance in the atmosphere. The United States should incorporate knowledge acquired from European projects and regulatory activities in addressing circular carbon economies and net-negative emissions.

FINDING 3.4 Life Cycle, Techno-economic, and Societal Factors Assessment. For all assessments of net-zero or net-negative emissions status of CO_2 utilization products, it is important to estimate the full life cycle impact of the process, including upstream and downstream greenhouse gas emissions associated with the process, feedstock origin, energy use, product fate, co-product fate, and associated waste. To ensure simultaneous economic viability and environmental justice, techno-economic assessments and societal factors have to be fully integrated with life cycle assessments.

> **RECOMMENDATION 3.2.** The U.S. Department of Energy (DOE) should build on ongoing efforts to harmonize and standardize life cycle assessment (LCA) for carbon capture, utilization, and storage projects. In addition to LCA, techno-economic assessment and analysis of societal impacts including environmental justice should be used by DOE to prioritize investments in project demonstrations and loan commitments.

FINDING 3.5 CO_2 Utilization Target Products. Targets for net-zero-emissions carbon products include concrete, aggregates, commodity carbon-containing chemicals, polymers, elemental carbon products, and high-value niche products. Many needs for carbon-based products will remain or grow in a net-zero-emissions economy, but a major class of materials, carbon-based fuels, is likely to decrease in use due to vehicle electrification. The use of zero-emission carbon fuels for heavy-duty transport, especially aviation, likely will remain substantial.

FINDING 3.6 Product Equivalence. Many CO_2-derived products (e.g., fuels, materials, chemicals) are chemically identical to their fossil-derived counterparts. These products can utilize existing infrastructure and systems and may displace incumbent fossil-derived products. Some CO_2-derived products, such as CO_2-cured concrete and carbonated aggregates, may have different composition, properties, and performance than the incumbent material, which could yield benefits or disadvantages.

FINDING 3.7 Displaced Emissions. Displacing fossil-derived products with CO_2 utilization-derived products may lead to avoidance of CO_2 emissions, if the CO_2 utilization processes are net-zero emissions. However, not all emissions may be displaced, particularly criteria emissions from combustion of fuels, such as nitrogen oxides and soot, which have local air quality and human health impacts.

FINDING 3.8 Net-Zero Emissions Enabling Inputs. For net-zero-emissions CO_2 utilization, all inputs—including but not limited to electricity, hydrogen, heat, and transportation—will need to have net-zero emissions on a life cycle basis. Electricity and hydrogen inputs can be substantially decarbonized by adding carbon capture and storage to existing infrastructure for generation from fossil fuels, or by building new infrastructure using zero-carbon-emissions options such as solar, wind, nuclear, or geothermal power.

FINDING 3.9 Energy and Hydrogen Requirements. Utilizing CO_2 as a starting material in chemical processes to produce fuels, commodity chemicals, and other hydrocarbon-based chemicals requires more external energy and often hydrogen inputs than generating the same products from fossil carbon sources. CO_2 utilization to produce hydrocarbon substitutes for fossil fuels will also require substantially more power and hydrogen than needed for direct use of electricity or hydrogen to power transportation and other energy services.

FINDING 3.10 Market Potential. The global utilization potential for CO_2 to make products is several gigatonnes per year. The volume of CO_2 utilized in a net-zero economy will be driven by the market value of carbon-based products and the competitiveness of CO_2 as a feedstock. These factors depend on (1) demand for services provided by carbon-based products; (2) their relative cost compared to fossil-based products, non-carbon-based alternatives such as electricity and hydrogen, and products derived from carbon sources other than CO_2, such as biomaterial carbon and recycled waste carbon products; (3) availability of inputs that enable net-zero product status, including clean hydrogen and energy; and (4) policy incentives and regulatory frameworks, including CO_2 abatement incentives. The potential for and cost of CO_2 removal will affect market demand directly in some cases—such as when the marketable product also provides carbon removal (e.g., concrete and aggregates or carbon fiber)—and indirectly in other cases, by changing the attractiveness of using CO_2 removal as an alternative to de-fossilization of chemical products.

3.8 REFERENCES

Adánez, J., and A. Abad. 2019. "Chemical-Looping Combustion: Status and Research Needs." *Proceedings of the Combustion Institute* 37(4):4303–4317. https://doi.org/10.1016/j.proci.2018.09.002.

Aether. n.d. "Diamonds from Thin Air." https://aetherdiamonds.com/?gclid=Cj0KCQjwg_iTBhDrARIsAD3Ib5iCN7jkzfw-7cO6eoWJNb8VOS9WJVb_p6U-wgnv3joI6XG-cIVTQHr0aApmYEALw_wcB.

Agora Verkehrswende, Agora Energiewende, and Frontier Economics. 2018. *The Future Cost of Electricity-Based Synthetic Fuels*. https://static.agora-energiewende.de/fileadmin/Projekte/2017/SynKost_2050/Agora_SynKost_Study_EN_WEB.pdf.

Air Protein. 2022. "How We'll Feed the Future." https://www.airprotein.com/making-air-meat.

AirMiners. 2022. "AirMiners Index." https://airminers.org/explore.

Arning, K., J. Offermann-van Heek, A. Sternberg, A. Bardow, and M. Ziefle. 2020. "Risk-Benefit Perceptions and Public Acceptance of Carbon Capture and Utilization." *Environmental Innovation and Societal Transitions* 35(June):292–308. https://doi.org/10.1016/j.eist.2019.05.003.

Arning, K., J. Offermann-van Heek, and M. Ziefle. 2021. "What Drives Public Acceptance of Sustainable CO_2-Derived Building Materials? A Conjoint-Analysis of Eco-Benefits vs. Health Concerns." *Renewable and Sustainable Energy Reviews* 144(July):110873. https://doi.org/10.1016/j.rser.2021.110873.

Artz, J., T.E. Müller, K. Thenert, J. Kleinekorte, R. Meys, A. Sternberg, A. Bardow, and W. Leitner. 2018. "Sustainable Conversion of Carbon Dioxide: An Integrated Review of Catalysis and Life Cycle Assessment." *Chemical Reviews* 118(2):434–504. https://doi.org/10.1021/acs.chemrev.7b00435.

Bazzanella, A.M., and F. Ausfelder. 2017. *Low Carbon Energy and Feedstock for the European Chemical Industry*. Technology Study. Germany: DECHEMA. https://dechema.de/dechema_media/Downloads/Positionspapiere/Technology_study_Low_carbon_energy_and_feedstock_for_the_European_chemical_industry-p-20002750.pdf.

Berg, M.C., D. Vural, S. Bhagia, L. Liang, Z. Yang, X. Meng, N. Gallego, N. Bryant, S.V. Pingali, and H.M. O'Neill. 2020. *Polymer and Structural Science Behind Valorizing Lignin Using Solvents*. Research Report. Oak Ridge National Laboratory. https://genomicscience.energy.gov/research-summaries-genomic-science-annual-pi-meeting-abstract-book-feb-2020.

Beswick, R.R., A.M. Oliveira, and Y. Yan. 2021. "Does the Green Hydrogen Economy Have a Water Problem?" *ACS Energy Letters* 6(9):3167–3169. https://doi.org/10.1021/acsenergylett.1c01375.

Biniek, K., K. Henderson, M. Rogers, and G. Santoni. 2020. "Driving CO_2 Emissions to Zero (and Beyond) with Carbon Capture, Use, and Storage." *McKinsey*, June 30. https://www.mckinsey.com/business-functions/sustainability/our-insights/driving-co2-emissions-to-zero-and-beyond-with-carbon-capture-use-and-storage.

Blanco, H. 2021." Hydrogen Production in 2050: How Much Water Will 74EJ Need?" *energypost.eu* July 22. https://energypost.eu/hydrogen-production-in-2050-how-much-water-will-74ej-need.

Buck, H.J. 2016. "Rapid Scale-up of Negative Emissions Technologies: Social Barriers and Social Implications." *Climatic Change* 139(2):155–167. https://doi.org/10.1007/s10584-016-1770-6.

CEQ (Council on Environmental Quality). 2022. "Carbon Capture, Utilization, and Sequestration Guidance." *Federal Register* 87(32):8808–8811.

chemeurope. n.d. "Methanol Economy." In *Encyclopedia of Chemistry*. https://www.chemeurope.com/en/encyclopedia/Methanol_economy.html.

Circular Carbon Network. 2022. "Innovator Index." https://circularcarbon.org/innovator-index.

Collodi, G., G. Azzaro, N. Ferrari, and S. Santos. 2017. "Techno-Economic Evaluation of Deploying CCS in SMR Based Merchant H_2 Production with NG as Feedstock and Fuel." *Energy Procedia* 114(July):2690–2712. https://doi.org/10.1016/j.egypro.2017.03.1533.

Cremonese, L., T. Strunge, B. Olfe-Kräutlein, S. Jahilo, T. Langhorst, S. McCord, L. Müller, et al. 2022. *Making Sense of Techno-Economic and Life Cycle Assessment Studies for CO_2 Utilization*. Ann Arbor, MI: Global CO_2 Initiative. https://doi.org/10.7302/4202.

CSIRO (Commonwealth Scientific and Industrial Research Organization). 2022. "CO_2 Utilisation Roadmap." https://www.csiro.au/en/work-with-us/services/consultancy-strategic-advice-services/CSIRO-futures/Energy-and-Resources/CO2-Utilisation-Roadmap.

DDP (Deep Decarbonization Pathways). n.d. "The DDP Initiative." https://ddpinitiative.org.

Dipple, G., P. Keleman, and C.M. Woodall. 2021. "The Building Blocks of CDR Systems." Chapter 2 in *Carbon Dioxide Removal Primer*, J. Wilcox, B. Kolosz, and J. Freeman, eds. CDR Primer. https://cdrprimer.org/read/chapter-2.

DOE-HFTO (U.S. Department of Energy Hydrogen and Fuel Cell Technologies Office). n.d. "DOE Technical Targets for Hydrogen Production from Electrolysis." https://www.energy.gov/eere/fuelcells/doe-technical-targets-hydrogen-production-electrolysis.

DOE-OCED (Office of Clean Energy Demonstrations). 2022. "DE-FOA-0002768: Notice of Intent to Issue Funding Opportunity Announcement No DE-FOA-0002779—Bipartisan Infrastructure Law: Additional Clean Hydrogen Programs (Section 40314): Regional Clean Hydrogen Hubs." https://oced-exchange.energy.gov/Default.aspx#FoaId4e674498-618c-4f1a-9013-1a1ce56e5bd3.

EC (European Commission). 2020. "A New Circular Economy Action Plan for a Cleaner and More Competitive Europe. Brussels, Belgium: EC." https://eur-lex.europa.eu/resource.html?uri=cellar:9903b325-6388-11ea-b735-01aa75ed71a1.0017.02/DOC_1&format=PDF.

Emergen Research. 2022. "Carbon Nanotube Market Size to Reach USD 3,027.4 Million in 2030: Increasing Application of Carbon Nanotubes in Electric Vehicles Is a Major Factor Driving Industry Demand, Says Emergen Research." *PR Newswire*, April 25. https://www.prnewswire.com/news-releases/carbon-nanotube-market-size-to-reach-usd-3-027-4-million-in-2030--increasing-application-of-carbon-nanotubes-in-electric-vehicles-is-a-major-factor-driving-industry-demand-says-emergen-research-301531992.html.

Energy Transitions Commission. 2018. *Mission Possible: Reaching Net-Zero Carbon Emissions from Harder-to-Abate Sectors by Mid-Century*. London.

Engelmann, L., K. Arning, A. Linzenich, and M. Ziefle. 2020. "Risk Assessment Regarding Perceived Toxicity and Acceptance of Carbon Dioxide-Based Fuel by Laypeople for Its Use in Road Traffic and Aviation." *Frontiers in Energy Research* 8(November):579814. https://doi.org/10.3389/fenrg.2020.579814.

Fernández-Dacosta, C., M. van der Spek, C.R. Hung, G.D. Oregionni, R. Skagestad, P. Parihar, D.T. Gokak, A.H. Strømman, and A. Ramirez. 2017. "Prospective Techno-Economic and Environmental Assessment of Carbon Capture at a Refinery and CO_2 Utilisation in Polyol Synthesis." *Journal of CO_2 Utilization* 21(October):405–422. https://doi.org/10.1016/j.jcou.2017.08.005.

Fortune Business Insights. 2021. *Graphite Market Size, Share & COVID-19 Impact Analysis, by Product (Synthetic, and Natural), by Application (Refractories, Foundries)*. Research Report. https://www.fortunebusinessinsights.com/graphite-market-105322.

Fortune Business Insights. 2022. *Graphene Market Size, Share & COVID-19 Impact Analysis, by Product (Graphene Oxide, Graphene Nanoplatelets (GNP), and Others), by End-Use Industry (Electronics, Aerospace & Defense, Energy, Automotive, and Others), and Regional Forecast, 2022-2029*. Research Report. https://www.fortunebusinessinsights.com/graphene-market-102930.

Friedlingstein, P., M.W. Jones, M. O'Sullivan, R.M. Andrew, D.C.E. Bakker, J. Hauck, C. Le Quéré, et al. 2022. "Global Carbon Budget 2021." *Earth System Science Data* 14(4):1917–2005. https://doi.org/10.5194/essd-14-1917-2022.

Gabrielli, P., M. Gazzani, and M. Mazzotti. 2020. "The Role of Carbon Capture and Utilization, Carbon Capture and Storage, and Biomass to Enable a Net-Zero-CO_2 Emissions Chemical Industry." *Industrial & Engineering Chemistry Research* 59(15):7033–7045. https://doi.org/10.1021/acs.iecr.9b06579.

Gallon, V. 2022. "Beiersdorf to Launch the World's First Skincare Product with Recycled CO_2." *Premium Beauty News* April 11. https://www.premiumbeautynews.com/en/beiersdorf-to-launch-the-world-s,20148.

Global Carbon Project. 2021. "Supplemental Data of Global Carbon Budget 2021 (Version 1.0) [Data set]." https://doi.org/10.18160/gcp-2021.

Global CO_2 Initiative. 2018. *Global Roadmap Study of CO_2U Technologies*. Technical Report. Ann Arbor: University of Michigan. https://hdl.handle.net/2027.42/146529.

Global CO_2 Initiative. 2022a. "AssessCCUS—Techno-Economic and Life Cycle Assessment for Carbon Capture, Utilization, and Storage." University of Michigan. https://assessccus.globalco2initiative.org.

Global CO_2 Initiative. 2022b. "CCU Activity Hub." University of Michigan. https://www.globalco2initiative.org/evaluation/carbon-capture-activity-hub.

Global Market Insights. 2021. *Carbon Black Market Size and Share: Statistics—2027*. Research Report. Selbyville, DE. https://www.gminsights.com/industry-analysis/carbon-black-market.

Haider, K. 2022. "CO_2-Derived Products." Panel discussion presentation at the Carbon Utilization Infrastructure, Markets, Research and Development Meeting #2. March 1. https://www.nationalacademies.org/event/03-01-2022/carbon-utilization-infrastructure-markets-research-and-development-meeting-2#sectionEventMaterials.

Hepburn, C., E. Adlen, J. Beddington, E.A. Carter, S. Fuss, N. Mac Dowell, J.C. Minx, P. Smith, and C. K. Williams. 2019. "The Technological and Economic Prospects for CO_2 Utilization and Removal." *Nature* 575(7781):87–97. https://doi.org/10.1038/s41586-019-1681-6.

Herzog, H.J. 2018. *Carbon Capture*. Cambridge, MA: MIT Press.

Hills, C.D., N. Tripathi, R.S. Singh, P.J. Carey, and F. Lowry. 2020. "Valorisation of Agricultural Biomass-Ash with CO_2." *Scientific Reports* 10(1):13801. https://doi.org/10.1038/s41598-020-70504-1.

Hopcroft, T., and A. Papadamou. 2021. "The Water Industry in a Hydrogen Economy." *The Water Report*, July 12. https://www.paconsulting.com/newsroom/expert-opinion/the-water-report-the-water-industry-in-a-hydrogen-economy-12-july-2021.

IEA (International Energy Agency). 2019. *Putting CO_2 to Use: Creating Value from Emissions*. Paris: IEA. https://iea.blob.core.windows.net/assets/50652405-26db-4c41-82dc-c23657893059/Putting_CO2_to_Use.pdf.

IEA. 2022. *Direct Air Capture 2022*. Technical Report. Paris. CC BY 4.0. https://www.iea.org/reports/direct-air-capture-2022.

IEAGHG (IEA Greenhouse Gas R&D Programme). 2017. *Techno-Economic Evaluation of SMF Based Standalone (Merchant) Hydrogen Plant with CCS*. Technical Report. Cheltenham, UK. https://ieaghg.org/publications/technical-reports.

IPCC (Intergovernmental Panel on Climate Change). 2022. *Climate Change 2022: Mitigation of Climate Change*. Working Group III Contribution to the Sixth Assessment Report of the Intergovernmental Panel on Climate Change. Cambridge, UK: Cambridge University Press https://report.ipcc.ch/ar6wg3/pdf/IPCC_AR6_WGIII_FinalDraft_FullReport.pdf.

IRENA (International Renewable Energy Agency) and Methanol Institute. 2021. *Innovation Outlook: Renewable Methanol*. Abu Dhabi: International Renewable Energy Agency. https://www.irena.org/-/media/Files/IRENA/Agency/Publication/2021/Jan/IRENA_Innovation_Renewable_Methanol_2021.pdf.

ITIF (Information Technology & Innovation Foundation). 2022. "Active Carbon Management: Critical Tools in the Climate Toolbox." https://itif.org/publications/2022/04/18/active-carbon-management-critical-tools-climate-toolbox.

Kahler, F., M. Carus, O. Porc, and C. vom Berg. 2021. *Turning Off the Tap for Fossil Carbon—Future Prospects for a Global Chemical and Derived Material Sector Based on Renewable Carbon*. Hurth, Germany: Nova Institute for Ecology and Innovation. https://renewable-carbon.eu/publications/product/turning-off-the-tap-for-fossil-carbon-future-prospects-for-a-global-chemical-and-derived-material-sector-based-on-renewable-carbon.

Kätelhön, A., R. Meys, S. Deutz, S. Suh, and A. Bardow. 2019. "Climate Change Mitigation Potential of Carbon Capture and Utilization in the Chemical Industry." *Proceedings of the National Academy of Sciences* 116(23):11187–11194. https://doi.org/10.1073/pnas.1821029116.

Kiverdi. 2019. "Carbon Transformation in Action." https://www.kiverdi.com/about.

Kondaveeti, S., I.M. Abu-Reesh, G. Mohanakrishna, M. Bulut, and D. Pant. 2020. "Advanced Routes of Biological and Bio-Electrocatalytic Carbon Dioxide (CO_2) Mitigation Toward Carbon Neutrality." *Frontiers in Energy Research* 8(June):94. https://doi.org/10.3389/fenrg.2020.00094.

Köpke, M., C. Mihalcea, J.C. Bromley, and S.D. Simpson. 2011. "Fermentative Production of Ethanol from Carbon Monoxide." *Current Opinion in Biotechnology* 22(3):320–325. https://doi.org/10.1016/j.copbio.2011.01.005.

La Plante, E.C., I. Mehdipour, I. Shortt, K. Yang, D. Simonetti, M. Bauchy, and G.N. Sant. 2021. "Controls on CO_2 Mineralization Using Natural and Industrial Alkaline Solids Under Ambient Conditions." *ACS Sustainable Chemistry & Engineering* 9(32):10727–10739. https://doi.org/10.1021/acssuschemeng.1c00838.

Lamb, J.J., M. Hillestad, E. Rytter, R. Bock, A.S.R. Nordgård, K.M. Lien, O.S. Burheim, and B.G. Pollet. 2020. "Chapter 3 – Traditional Routes for Hydrogen Production and Carbon Conversion." Pp. 21–53 in *Hydrogen, Biomass and Bioenergy*, J.J. Lamb and B.G. Pollet, eds. Cambridge, MA: Academic Press. https://doi.org/10.1016/B978-0-08-102629-8.00003-7.

Lampert, D.J., H. Cai, Z. Wang, J. Keisman, M. Wu, J. Han, J. Dunn, et al. 2015. *Development of a Life Cycle Inventory of Water Consumption Associated with the Production of Transportation Fuels*. ANL/ESD-15/27. Argonne, IL: Argonne National Laboratory. https://doi.org/10.2172/1224980.

Lampert, D.J., H. Cai, and A. Elgowainy. 2016. "Wells to Wheels: Water Consumption for Transportation Fuels in the United States." *Energy & Environmental Science* 9(3):787–802. https://doi.org/10.1039/C5EE03254G.

Langanke, J., A. Wolf, J. Hofmann, K. Böhm, M.A. Subhani, T.E. Müller, W. Leitner, and C. Gürtler. 2014. "Carbon Dioxide (CO_2) as Sustainable Feedstock for Polyurethane Production." *Green Chemistry* 16(4):865–870. https://doi.org/10.1039/C3GC41788C.

Lange, J.-P. 2021. "Towards Circular Carbo-Chemicals—the Metamorphosis of Petrochemicals." *Energy & Environmental Science* 14(8):4358–4376. https://doi.org/10.1039/D1EE00532D.

Larson, E., C. Greig, J. Jenkins, E. Mayfield, A. Pascale, C. Zhang, J. Drossman, et al. 2021. *Net-Zero America: Potential Pathways, Infrastructure, and Impacts*. Carbon Mitigation Initiative at Princeton University. Princeton, NJ. https://netzeroamerica.princeton.edu/the-report.

Li, V.C. 2019. *Engineered Cementitious Composites (ECC): Bendable Concrete for Sustainable and Resilient Infrastructure*. Berlin, Heidelberg: Springer. https://doi.org/10.1007/978-3-662-58438-5.

Liu, N. 2021. "Increasing Blue Hydrogen Production Affordability." *Hydrocarbon Processing* June. https://www.hydrocarbonprocessing.com/magazine/2021/june-2021/special-focus-process-optimization/increasing-blue-hydrogen-production-affordability.

Luque, R.L., and J.G. Speight. 2015. *Gasification for Synthetic Fuel Production: Fundamentals, Processes, and Applications*. Cambridge, UK: Woodhead Publishing. https://doi.org/10.1016/C2013-0-16368-4.

Lutzke, L., and J. Árvai. 2021. "Consumer Acceptance of Products from Carbon Capture and Utilization." *Climatic Change* 166(1–2):15. https://doi.org/10.1007/s10584-021-03110-3.

Lux Research. 2020. *CO_2 Capture & Utilization: The Emergence of a Carbon Economy*. State of the Market Report. Boston, MA: Lux Research. https://members.luxresearchinc.com/research/report/35957.

Mac Dowell, N., P.S. Fennell, N. Shah, and G.C. Maitland. 2017. "The Role of CO_2 Capture and Utilization in Mitigating Climate Change." *Nature Climate Change* 7:243–247.

Mangan, A. 2010. "By-Product Synergy Networks: Driving Innovation Through Waste Reduction and Carbon Mitigation." Chapter 6 in *Sustainable Development in the Process Industries: Cases and Impact*, J. Harmsen and J.B. Powell, eds. New York: John Wiley & Sons.

Marchese, M., E. Giglio, M. Santarelli, and A. Lanzini. 2020. "Energy Performance of Power-to-Liquid Applications Integrating Biogas Upgrading, Reverse Water Gas Shift, Solid Oxide Electrolysis and Fischer-Tropsch Technologies." *Energy Conversion and Management: X* 6(April):100041. https://doi.org/10.1016/j.ecmx.2020.100041.

McCoy, M. 2022. "Green Chemical Maker LanzaTech to Go Public via Merger." *Chemical & Engineering News* March 9. https://cen.acs.org/business/biobased-chemicals/Green-chemical-maker-LanzaTech-to-go-public-via-merger/100/web/2022/03.

Monkman, S., and M. MacDonald. 2017. "On Carbon Dioxide Utilization as a Means to Improve the Sustainability of Ready-Mixed Concrete." *Journal of Cleaner Production* 167(November):365–375. https://doi.org/10.1016/j.jclepro.2017.08.194.

Mordor Intelligence. 2021. "Quantum Dots Market – Growth, Trends, COVID-19 Impact, and Forecasts (2022-2027)." Industry Forecast. https://www.mordorintelligence.com/industry-reports/quantum-dots-market-industry.

Müller, L.J., A. Kätelhön, S. Bringezu, S. McCoy, S. Suh, R. Edwards, V. Sick, et al. 2020. "The Carbon Footprint of the Carbon Feedstock CO_2." *Energy & Environmental Science* 13(9):2979–2992. https://doi.org/10.1039/D0EE01530J.

NASEM (National Academies of Sciences, Engineering, and Medicine). 2019. *Gaseous Carbon Waste Streams Utilization: Status and Research Needs*. Washington, DC: The National Academies Press. https://doi.org/10.17226/25232.

NASEM. 2021a. *Accelerating Decarbonization of the U.S. Energy System*. Washington, DC: The National Academies Press. https://doi.org/10.17226/25932.

NASEM. 2021b. *Assessment of Technologies for Improving Light-Duty Vehicle Fuel Economy—2025–2035*. Washington, DC: The National Academies Press. https://doi.org/10.17226/26092.

NPC (National Petroleum Council). 2019. *Meeting the Dual Challenge: A Roadmap to At-Scale Deployment of Carbon Capture, Use, and Storage*. Technical Report. Washington, DC. https://dualchallenge.npc.org.

Oni, A.O., K. Anaya, T. Giwa, G. Di Lullo, and A. Kumar. 2022. "Comparative Assessment of Blue Hydrogen from Steam Methane Reforming, Autothermal Reforming, and Natural Gas Decomposition Technologies for Natural Gas-Producing Regions." *Energy Conversion and Management* 254(February):115245. https://doi.org/10.1016/j.enconman.2022.115245.

Pace, G., and S.W. Sheehan. 2021. "Scaling CO_2 Capture with Downstream Flow CO_2 Conversion to Ethanol." *Frontiers in Climate* 3(May):656108. https://doi.org/10.3389/fclim.2021.656108.

Poland, C. 2021. "The Air Force Partners with Twelve, Proves It's Possible to Make Jet Fuel Out of Thin Air." Air Force, October 22. https://www.af.mil/News/Article-Display/Article/2819999/the-air-force-partners-with-twelve-proves-its-possible-to-make-jet-fuel-out-of.

Ravikumar, D., G.A. Keoleian, S.A. Miller, and V. Sick. 2021a. "Assessing the Relative Climate Impact of Carbon Utilization for Concrete, Chemical, and Mineral Production." *Environmental Science & Technology* 55(17):12019–12031. https://doi.org/10.1021/acs.est.1c01109.

Ravikumar, D., D. Zhang, G. Keoleian, S. Miller, V. Sick, and V. Li. 2021b. "Carbon Dioxide Utilization in Concrete Curing or Mixing Might Not Produce a Net Climate Benefit." *Nature Communications* 12(1):855. https://doi.org/10.1038/s41467-021-21148-w.

Research and Markets. 2020. "Polymers Market—Forecast (2020–2025)." https://www.researchandmarkets.com/reports/5021563/polymers-market-forecast-2020-2025.

Research Reports World. 2021a. "Global Amorphous Graphite Market Research Report: Growth Trends and Competitive Analysis 2021–2027." https://www.researchreportsworld.com/global-amorphous-graphite-industry-18347737.

Research Reports World. 2021b. "Global Carbon Carbon Composites Industry Research Report: Growth Trends and Competitive Analysis 2021–2027." https://www.researchreportsworld.com/global-carbon-carbon-composites-industry-18344501.

Searle, S., G. Bieker, and C. Baldino. 2021. *Decarbonizing Road Transport by 2050: Zero-Emission Pathways for Passenger Vehicles*. Briefing Paper. Washington, DC: International Council on Clean Transportation. https://theicct.org/wp-content/uploads/2021/12/zevtc-decarbonizing-by-2050-Jul2021%E2%80%AF.pdf.

Sick, V., G. Stokes, and F.C. Mason. 2022a. "CO_2 Utilization and Market Size Projection for CO_2-Treated Construction Materials." *Frontiers in Climate* 4(May):878756. https://doi.org/10.3389/fclim.2022.878756.

Sick, V., G. Stokes, F. Mason, Y.-S. Yu, A. Van Berkel, R. Daliah, O. Gamez, C. Gee, and M. Kaushik. 2022b. *Implementing CO_2 Capture and Utilization at Scale and Speed: The Path to Achieving Its Potential*. Technical Report. Global CO_2 Initiative, University of Michigan. https://dx.doi.org/10.7302/5825.

Tenhumberg, N., and K. Büker. 2020. "Ecological and Economic Evaluation of Hydrogen Production by Different Water Electrolysis Technologies." *Chemie Ingenieur Technik* 92(10):1586–1595. https://doi.org/10.1002/cite.202000090.

Third Derivative. n.d. "First Gigaton Captured." https://www.third-derivative.org/first-gigaton-captured.

Timmerberg, S., M. Kaltschmitt, and M. Finkbeiner. 2020. "Hydrogen and Hydrogen-Derived Fuels Through Methane Decomposition of Natural Gas—GHG Emissions and Costs." *Energy Conversion and Management: X* 7(September):100043. https://doi.org/10.1016/j.ecmx.2020.100043.

Troschl, C., K. Meixner, and B. Drosg. 2017. "Cyanobacterial PHA Production—Review of Recent Advances and a Summary of Three Years' Working Experience Running a Pilot Plant." *Bioengineering* 4(2):26. https://doi.org/10.3390/bioengineering4020026.

UNFCCC (United Nations Framework Convention on Climate Change). n.d. "Common Metrics." https://unfccc.int/process-and-meetings/transparency-and-reporting/methods-for-climate-change-transparency/common-metrics.

Uygun, H.D.E., and Z.O. Uygun. 2020. "Fullerene Based Sensor and Biosensor Technologies." Chapter 4 in *Smart Nanosystems for Biomedicine, Optoelectronics and Catalysis*. Rijeka: IntechOpen. https://doi.org/10.5772/intechopen.93316.

Wang, Z., T. Hu, R. Liang, and M. Wei. 2020. "Application of Zero-Dimensional Nanomaterials in Biosensing." *Frontiers in Chemistry* 8. https://doi.org/10.3389/fchem.2020.00320.

Wolske, K.S., K.T. Raimi, V. Campbell-Arvai, and P.S. Hart. 2019. "Public Support for Carbon Dioxide Removal Strategies: The Role of Tampering with Nature Perceptions." *Climatic Change* 152:345–361. https://doi.org/10.1007/s10584-019-02375-z.

Woodall, C.M., N. McQueen, H. Pilorgé, and J. Wilcox. 2019. "Utilization of Mineral Carbonation Products: Current State and Potential." *Greenhouse Gases: Science and Technology* 9(6):1096–1113. https://doi.org/10.1002/ghg.1940.

Zhang, D., T. Liu, and Y. Shao. 2020. "Weathering Carbonation Behavior of Concrete Subject to Early-Age Carbonation Curing." *Journal of Materials in Civil Engineering* 32(4):04020038. https://doi.org/10.1061/(ASCE)MT.1943-5533.0003087.

Zimmermann, A., L. Müller, A. Marxen, K. Armstrong, G. Buchner, J. Wunderlich, S. Michailos, et al. 2018. *Techno-Economic Assessment & Life Cycle Assessment Guidelines for CO_2 Utilization*. Global CO_2 Initiative, University of Michigan. https://doi.org/10.3998/2027.42/145436.

Zimmerman, J.B., P.T. Anastas, H.C. Erythropel, and W. Leitner. 2020. "Designing for a Green Chemistry Future." *Science* 367(6476):397–400. https://doi.org/10.1126/science.aay3060.

4

Infrastructure Considerations for CO_2 Utilization

This chapter describes considerations for developing infrastructure for carbon dioxide (CO_2) utilization, taking into account the CO_2-derived products identified in Chapter 3 and the existing infrastructure discussed in Chapter 2. Infrastructure needs throughout the CO_2 utilization value chain are examined, from capture to purification, transportation, conversion, and, where applicable, transportation of the CO_2-derived product. Requirements for enabling infrastructure, namely, clean electricity, hydrogen, water, land, and energy storage, are also considered.

4.1 CO_2 CAPTURE

There are currently only 12 commercial CO_2 capture and storage (CCS) facilities in the United States, with a combined installed capacity of around 20 million tonnes per annum (Mtpa), and two more plants under construction (Global CCS Institute 2022). As noted in Chapter 2, 11 of these CCS facilities capture CO_2 for use in enhanced oil recovery (EOR), with only one having geologic CO_2 storage as its sole purpose. The Petra Nova project was the only large U.S. commercial electricity-generating plant to be equipped with carbon capture equipment, and that project was shut down in 2020 due to a combination of the economic effects of the COVID-19 pandemic and persistently low oil prices that reduced the economic benefit of using captured CO_2 for EOR (Anchondo and Klump 2020). Per the Global CCS Institute's CO_2RE database, as of June 2022, about 60 carbon capture projects are in various stages of development in the United States, only one of which is for direct air capture (DAC) (Global CCS Institute 2022). The Clean Air Task Force maintains a database of publicly announced CO_2 capture, utilization, and storage (CCUS) projects under development in the United States, which includes information about the project status (e.g., announced, pre-front-end engineering design [pre-FEED], FEED, FEED complete), location, capture capacity, and sector (CATF 2022). Figure 4-1 shows the geographic locations, sectors, and capture capacities of these projects.

Carbon capture technologies are at varying levels of maturity today. Amine-based solvents for post-combustion capture constitute the most advanced carbon capture technology and are the most commonly deployed in commercial carbon capture facilities. However, other technologies have been deployed commercially in the United States, such as vacuum swing adsorption (e.g., at the Air Products' Port Arthur Hydrogen CCS project in Texas) and cryogenic separation performed at Exxon's Shute Creek natural gas processing facility in Wyoming. Other technologies including sorbents, membranes, and chemical and calcium looping cycles are in various stages of development.

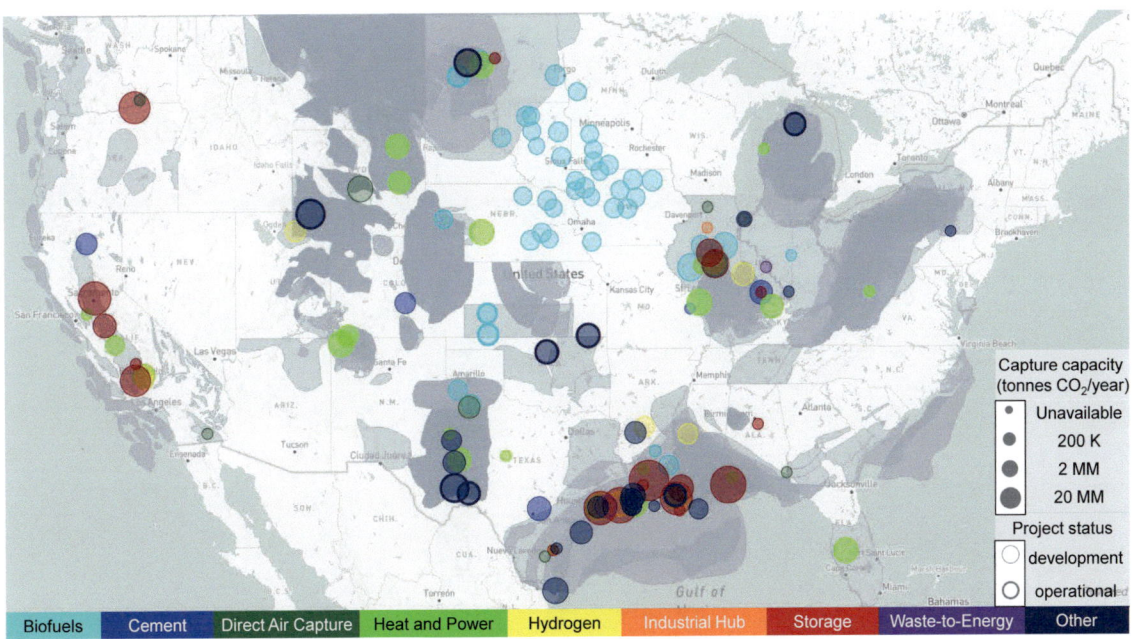

FIGURE 4-1 Map of carbon capture projects announced or under development in the United States color-coded by sector, where the circle size represents capture capacity, ranging from 200,000 to 20 million tonnes CO_2 per year. Dark gray shading represents potential locations for geologic storage of CO_2. No carbon capture projects are under development in Alaska or Hawaii.
SOURCE: Adapted from Clean Air Task Force, 2021, "US Carbon Capture Activity and Project Map," https://www.catf.us/ccsmapus.

Multiple assessments, including those by the International Energy Agency (IEA 2020), Concawe (2021), and the Global CCS Institute (Kearns et al. 2021), analyze the maturity of different carbon capture technologies. Drawing from these publications, Table 4-1 summarizes the maturity of select carbon capture technologies.

Today's commercial CCS facilities are deployed at natural gas processing plants, fertilizer plants, bioethanol plants, coal-fired power plants, and hydrogen production facilities, while other applications are under development. In practice, the most appropriate capture technology for a given application depends on several factors, including the initial and final desired CO_2 concentration (i.e., the percentage of CO_2 to be removed), scale of CO_2 capture, operating pressure and temperature, composition and flow rate of the gas stream, integration with the original facility, and cost considerations.

Post-combustion capture technologies can be retrofitted at the exhaust point of existing combustion and other CO_2-emitting facilities. New combustion facilities employing oxy-fuel combustion and pre-combustion CO_2 capture provide alternative approaches that ultimately may result in lower capture costs due to higher concentrations of CO_2. Oxy-combustion power production uses oxygen instead of air for fuel combustion, resulting in a flue gas stream that is approximately 70 percent CO_2 by volume with the balance being water vapor, oxygen, and trace compounds. Oxy-combustion CO_2 capture is applicable only to facilities with a special oxy-combustion boiler, and requires significant capital investment in an air separation unit for oxygen production in addition to CO_2 purification equipment. Oxy-combustion can have broader application in industry, such as for metals and glass production (Linde 2022), where oxygen enhancement both improves process efficiency and reduces fuel consumption and its associated greenhouse gas (GHG) and other noxious emissions. In pre-combustion capture processes, a fuel is transformed into hydrogen so that the CO_2 can be removed from the feedstock prior to combustion. Solid or gaseous fuels can be converted first to synthesis gas (syngas, carbon monoxide and hydrogen) via either gasification or partial oxidation (PO_x) under high pressure and temperature. This syngas can undergo the

TABLE 4-1 Technology Readiness Levels of Select Carbon Capture Technologies

Carbon Capture Technology	Technology Status (Technology Readiness Level)			
	Emerging (1–3)	Under Development (4–6)	Near Commercial (7–8)	Commercial (9)
Solvents	• Encapsulated solvents • Ionic liquids	• Water-lean solvents • Nonaqueous solvents • Phase-change solvents • Amino acid–based solvents • Biphasic solvents • Aminosilicone solvents	• Chilled ammonia • Mixed-salt solvent • Activated methyldiethanolamine solution • Piperazine	• Traditional amine solvents • Benfield process and variants • Physical solvents (Selexol, Rectisol) • Sterically hindered amines
Sorbents	• Electrochemically mediated adsorption	• Enzyme-catalyzed adsorption • Sorbent-enhanced water-gas shift • Alkalized alumina sorbent	• Rapid-temperature-swing adsorption • Metal organic frameworks	• Solid amine sorbents • Pressure swing adsorption (PSA) • Vacuum swing adsorption (VSA)
Membranes	• Room-temperature ionic liquid membranes • Catalytic membrane reactor • Electrodialysis • Membrane contactors	• Polymeric membranes/cryogenic separation hybrid • Polymeric membranes/solvent hybrid • Ceramic membrane • Zeolite membrane • PEEK membrane	• Polymeric membranes • Electrochemical membrane integrated with molten carbonate fuel cells	• Natural gas processing membranes
Other technologies	• Hydrolytic softening	• Calcium looping • Chemical combustion looping • Allam-Fetvedt cycle • Calix advanced calciner		

SOURCES: Data from Concawe (2021); IEA (2020); Kearns et al. (2021).

water-gas shift reaction to convert CO and added steam to hydrogen and a CO_2-rich gas. Autothermal reforming (ATR), an alternative hydrogen production process, combines the steam reforming and partial oxidation steps into a single reactor, followed by a series of shift reactions to generate syngas with CO_2 concentration of 25–30 percent by volume. ATR was developed in the 1950s and is used in commercial applications to provide syngas for ammonia and methanol synthesis but has not been deployed commercially yet for hydrogen production with carbon capture. Since syngas is at high pressure (~30 bar), the CO_2 can be cost-effectively separated. It is also possible to produce hydrogen with steam methane reforming (SMR), where CO_2 is captured from the flue gas (18–20 percent by volume, atmospheric pressure).

The cost of CO_2 capture can vary significantly but is inversely proportional to the concentration of CO_2 in the gas stream, with CO_2 capture from ambient air at ~420 parts per million by volume (0.042% CO_2 in air) being the costliest of all. The plant's location, energy and steam supply, and integration with the original facility will also have an impact on the cost of capture (Ferrari et al. 2019; IEAGHG 2018). Though CO_2 capture is implemented, its cost has been a deterrent to more widespread deployment. The cost of CO_2 capture today can be as high as $600/tonne captured CO_2 in the case of DAC (Tollefson 2018) or as low as $15/tonne in the case of industrial CO_2 capture from natural gas processing plants where CO_2 concentrations are above 95 percent (IEA 2021). DAC has seen a surge in interest and investment over the past few years, and a growing number of companies are entering the space. Nonetheless, achieving the U.S. Department of Energy's (DOE's) target cost of less than $100/net tonne of CO_2-equivalent by 2030 will require significant research and development (R&D). Ongoing DOE financial support for CO_2 capture FEED studies and large pilot or demonstration facilities may spur additional investment in commercial deployments. For example, DOE recently released a notice of intent for a Carbon Capture Demonstration Projects Program to fund projects that "will address technical, environmental, permitting, and financing challenges for commercial deployment" (DOE 2022). Table 4-2 summarizes some of the necessary improvements in next-generation innovations to drive down the costs and improve the performance of technologies for CO_2 capture.

TABLE 4-2 RD&D Targets to Improve Carbon Capture Systems

CO_2 Capture Technology	Research Trends for Reducing Carbon Capture Costs
Advanced solvents	• Fast sorption and desorption kinetics • Lower regeneration energy requirements • Lower degradation rates • Water-lean solvents
Sorbents	• Low-cost materials with high CO_2 adsorption rate and capacity • Fast spent sorbent regeneration rates • Improved durability over multiple regeneration cycles with little to no attrition • Low heats of adsorption • Adequately hydrophobic
Membranes	• High CO_2 permeability and selectivity • Low-cost materials • Improved durability determined by mechanical strength, chemical resistance, and thermal stability • Integration into low-pressure drop modules • Highly hydrophobic
Novel concepts	• Electrochemical capture • Crystallization • Microwave enhancement

SOURCE: Data from National Energy Technology Laboratory, 2020, *2020 Carbon Capture Program R&D: Compendium of Carbon Capture Technology*, Pittsburgh: National Energy Technology Laboratory, https://www.netl.doe.gov/sites/default/files/2020-07/Carbon-Capture-Technology-Compendium-2020.pdf.

Because of the additional energy required to capture CO_2, deployment of carbon capture technology can impact emissions of other pollutants. It is important to differentiate between emissions at the life cycle level from direct emissions. At the location of the capture unit, emissions of nitrogen oxides (NO_x), sulfur oxides (SO_x), and particulate matter are likely to decrease because of the additional purification of the flue gas stream required before it enters the capture unit. Most carbon capture technologies are poisoned by sulfur compounds, so flue gas pre-treatment for carbon capture removes SO_x in excess of what is scrubbed out during typical flue gas desulfurization in power plants (EEA 2020). On the other hand, capture technologies may increase emissions of some pollutants; for instance, commercially available amine-based capture systems may cause local increases in volatile organic compounds due to the carryover of compounds from solvent degradation in the treated flue gas (NETL 2020). Emissions of ammonia (NH_3) due to amine solvent degradation or nitrosamine formation from reaction of the amine solvent with NO_x in the flue gas can also occur (Benquet et al. 2021; Gibbins and Lucquiaud 2021; Gorset et al. 2014; Spietz et al. 2017). Because solvent losses from post-combustion capture—whether due to degradation products such as nitrosamines or solvent carryover—have economic impacts as well as potential health and environmental effects, their mitigation is an active area of research. Some system designs can integrate CO_2 capture synergistically with removal of SO_x and NO_x in order to reduce criteria pollutants in addition to CO_2 (Shaw 2009). Flue gas pre-treatments to reduce aerosol concentrations, and as a result minimize solvent losses, are also being explored (see, e.g., Bostick 2019, 2022). Oxy-combustion reduces air pollutants because it does not heat and react with the nitrogen component of air, so there are significantly fewer NO_x emissions compared to conventional combustion processes.

At the life cycle level, the impact on non-CO_2 emissions depends on the fuel used in the capture unit and its effect on the system. For instance, in a power plant, the heat required for regenerating the solvent in post-combustion capture is generally extracted from the turbine of the power plant, resulting in an efficiency penalty (i.e., less net electricity output per unit of fuel input). If this penalty is compensated for using the same type of power plant, a larger amount of fuel would be needed to produce the same amount of electricity output, resulting in an increase in life cycle emissions of air pollutants (Corsten et al. 2013).

4.2 CO_2 PURIFICATION

Depending on the origin of the CO_2 source, the CO_2 flow contains different impurities, which can impact CO_2 transport, utilization, and storage. Impurities such as water, sulfur trioxide (SO_3), nitrogen dioxide (NO_2), hydrogen sulfide (H_2S), and oxygen can affect corrosion rates. Nitrogen, argon, and hydrogen reduce fluid density and can increase compression costs. Impurities also can affect the phase boundaries and make it more difficult to maintain single-phase flow. The maximum concentration of toxic or flammable gases such as H_2S, carbon monoxide, and hydrocarbons is not determined by technical constraints, but rather by risk assessments of potential pipeline failure, which will depend on regional aspects such as population density. Table 4-3 shows an overview of impurities in CO_2 flows by type of source. These impurities change depending on the specific component in the feedstock, the type of process and process specifications, and the capture process. Limited literature exists on trace impurities (an overview is shown in Table 4-4), but these need to be considered when designing CO_2 capture and utilization (CCU) processes. Table 4-5 gives the maximum impurity thresholds by transport type as reported in different studies and projects, showing that the thresholds can differ significantly among regions. In countries such as Norway and the United Kingdom, where large-scale CO_2 transport (for CCS) has just begun to be deployed, the limits are stricter. There are trade-offs between the cost of cleaning the flue gas and the operational costs and lifetime of the pipelines.

The required purity of CO_2 feedstock varies with the utilization process, and Table 4-6 gives an overview of impurities of concern by utilization type. The potential impact of impurities on CO_2 utilization has not been studied extensively. In fact, most studies generally assume that the CO_2 feed to the utilization process is 100 percent pure. As indicated in Table 4-5, this will not be the case; even with strict CO_2 transport thresholds, there still will be impurities that can affect the utilization processes. For example, although biological CO_2 conversion is fairly impurity tolerant, care must be taken to avoid oxygen if the processes require an anaerobic environment. On the other end of the spectrum, electrochemical CO_2 conversion requires high-purity CO_2 streams and therefore may

TABLE 4-3 Overview of Impurity Concentrations of CO_2 Streams from Different Illustrative Facility Types

Component	Subcritical Pulverized Bituminous Coal (Illinois #6) Plant with Post-Combustion Capture[a] Gas leaving the carbon capture unit (post combustion with MEA[b])	Natural Gas with Carbon Capture[c] Gas leaving the carbon capture unit (post combustion with MEA[b])	Oxyfuel Combustion at Supercritical Pulverized Coal Plant[a,d] Gas leaving the boiler unit	Cement Plant[a] Gas leaving the carbon capture unit (post combustion with MEA[b])	Refinery Stack[a] Gas leaving the carbon capture unit (post combustion with MEA[b])	Bioethanol Plant[e] Raw CO_2 gas from ethanol plant	Direct Air Capture[f] Gas leaving the capture unit (KOH sorbent)
CO_2	99.7%	95%	96.65%	99.8%	99.6%	90%	97.11%
CO			750 ppmv	1.2 ppmv			
H_2O	640 ppmv	4%	100 ppmv	640 ppmv	640 ppmv	1–5 ppmv	0.01%
CH_4				0.026 ppmv		0–3 ppmv	
SO_2	<1 ppmv		50 ppmv	<0.1 ppmv	1.3 ppmv		
SO_3			20 ppmv				
NO_2	1.5 ppmv			0.86 ppmv	2.5 ppmv		
NO_x			100 ppmv				
O_2	61 ppmv		0.81%	35 ppmv	121 ppmv	10–100 ppmv	1.36%
H_2S		200 ppmv			7.9 ppmv		
N_2	0.18%	0.5%	1.96%	893 ppmv	0.29%	50–600 ppmv	1.51%
Ar	22 ppmv		0.57%	11 ppmv	38 ppmv		
Hg	0.0007 ppmv		0.011 ppmv	0.00073 ppmv			
As	0.0055 ppmv		0.026 ppmv	0.0029 ppmv			
Se	0.017 ppmv		0.08 ppmv	0.0088 ppmv			
Cl	0.85 ppmv			0.41 ppmv	0.4 ppmv		
Ethanol						25–950 ppmv	
Methanol						1–50 ppmv	
Acetaldehyde						3–75 ppmv	
Isoamyl acetate						0.6–3.0 ppmv	
Isobutanol						0–3 ppmv	
Ethylacetate						2–30 ppmv	

[a] Values from EC (2011).
[b] MEA = monoethanolamine
[c] Values from SINTEF (2019).
[d] Values from Rütters et al. (2015).
[e] Values from McKaskle et al. (2018).
[f] Values from Keith et al. (2018).

TABLE 4-4 Overview of Trace Impurities by CO_2 Source

Impurity	Combustion	Wells/Geothermal	Fermentation/Bioethanol Anaerobic Digestion (purely energy crops)	Anaerobic Digestion (waste)	Hydrogen or Ammonia	Phosphate Rock	Coal Gasification	Ethylene Oxide	Acid Neutralization	Vinyl Acetate
Aldehydes	✓	✓	✓	✓	✓		✓	✓		✓
Amines	✓				✓					
Benzene	✓	✓	✓	✓	✓		✓	✓	✓	✓
Carbon monoxide	✓	✓	✓	✓	✓	✓	✓	✓	✓	✓
Carbonyl sulfide	✓	✓	✓	✓	✓		✓	✓		✓
Cyclic aliphatic hydrocarbons	✓	✓		✓	✓		✓	✓		✓
Dimethyl sulfide		✓	✓	✓		✓	✓		✓	
Ethanol	✓	✓	✓	✓	✓		✓	✓		✓
Ethers		✓	✓	✓	✓		✓	✓		✓
Ethyl acetate		✓	✓	✓			✓	✓		✓
Ethyl benzene		✓		✓	✓		✓	✓		✓
Ethylene oxide								✓		✓
Halocarbons	✓			✓			✓	✓		✓
Hydrogen cyanide	✓						✓			
Hydrogen sulfide	✓	✓	✓	✓	✓	✓	✓	✓	✓	✓
Ketones	✓	✓	✓	✓	✓		✓	✓		✓
Mercaptans	✓	✓	✓	✓	✓	✓	✓	✓		✓
Mercury	✓	✓					✓			
Methanol	✓		✓	✓	✓		✓	✓		✓
Nitrogen oxides	✓		✓	✓	✓		✓	✓	✓	
Phosphine						✓				
Radon		✓				✓			✓	
Sulfur dioxide	✓	✓	✓	✓	✓	✓	✓		✓	
Toluene		✓	✓	✓	✓		✓	✓		✓
Vinyl chloride	✓						✓	✓		✓
Volatile hydrocarbons	✓	✓	✓	✓	✓		✓	✓		✓
Xylene		✓	✓	✓	✓		✓	✓		✓

SOURCE: Adapted from European Industrial Gases Association, 2016, "Carbon Dioxide Food and Beverages Grade, Source Qualification, Quality Standards and Verification," EIGA Doc 70/17, revision of Doc 70/08, https://www.eiga.eu/ct_documents/doc070-pdf.

TABLE 4-5 Overview of Recommended Maximum Impurity Limits for CO_2 Transport in Pipelines and Shipping

Component	Pipelines		National Grid Carbon (United Kingdom)[b]	Northern Lights Project (Norway)[c]	Shipping
	NETL (United States)[a]				EU CCUS Projects Network[d]
	Conceptual Design	Range in Literature			
H_2O (ppmv)	500	20–650	50	30	50
N_2 vol%	4	0–7			<0.3
O_2 vol%	0.001	0.001–4	0.001	10	Unknown
Ar vol%	4	0.01–4			<0.3
CH_4 vol%	4	0.01–4			<0.3
H_2 vol%	4	0–4	2	50	<0.3
CO ppmv	35	10–5,000	200	100	2,000
H_2S vol%	0.01	0.002–1.3	0.002 (for dense-phase 150 barg) 0.008 (for gas-phase 38 barg)	0.0009	200 ppm
SO_2 ppmv	100	10–50,000			
SO_x ppmv			100		
NO_x ppmv	100	20–2,500	100	10	
NH_3 ppmv	50	0–50		10	
COS ppmv	trace	trace			
C_2H_6 ppmv	1	0–1			
C_{3+} ppmv	<1	0–1			
Particulates ppmv	1	0–1			
Hg ppmv				0.03	
Glycol ppmv	46	0–174			
Cd, Tl, ppm				0.03 (sum)	

[a] Values from NETL (2019).
[b] Values from Gibbins and Lucquiaud (2021).
[c] Values from Equinor (2019).
[d] Values from Brownsort (2019).

be most feasible if paired with DAC CO_2 because of the latter's limited contamination. Indeed, the number of explorative CO_2 utilization studies that use DAC as the CO_2 source is increasing in the literature. Further purification of CO_2 streams can be accomplished using existing technologies (see, e.g., Abbas et al. 2013), so this does not necessarily present a technological challenge but rather an issue to consider when selecting the source of the CO_2 and designing CO_2 utilization routes, since further purifying CO_2 flows to desired levels significantly impacts CO_2 capture and utilization costs. Given that no single purity specification would be appropriate for all processes, infrastructure decisions for CO_2 transport likely will be based on purity needed for transport, with any further purification performed at the utilization facility if necessary.

TABLE 4-6 Overview of Impurities of Concern by CO_2 Utilization Route

CO_2 Utilization Route	Required Purity	Impurities of Concern
Mineralization	Low	Most processes can work directly with flue gas if desired
Biological conversion (anaerobic)	Low to medium	High tolerance to impurities except for oxygen
Thermochemical conversion	High to very high	Metals, sulfur can damage the catalyst
Electrochemical conversion	Very high	Metals, sulfur (SO_2, H_2S, COS) at part-per-million levels can damage the electrochemical reactor
CO_2 with food-grade purity[a]	Very high	Carbon monoxide, hydrocarbons, metals (CO_2: >99%; H_2O: <2 ppm; CO: <10 ppm; C_xH_y: < 50 ppmv; oil: <10 ppmw; and passing tests for acidity and red substances)

[a] Although use of CO_2 in the food and beverage industry is out of scope for this report, these purity levels have been included as a point of comparison, since many CO_2 utilization studies assume that the input CO_2 is food-grade purity. Purity levels are taken from EIGA (2018).

4.3 CO_2 TRANSPORTATION

CO_2 capture from point sources, the atmosphere, or bodies of water can be transported using modular means (ship, barges, trucks, or rail), high-pressure pipelines, or most likely, a combination of these various means, especially in the case of some stranded emitters where the construction of stand-alone conversion facilities may not be economically viable. Of the various modes of transportation, pressurized pipelines are considered by far the most cost-effective way of transporting large volumes of CO_2 in the supercritical or dense phase.

As discussed in Chapter 2, the United States has more than 5,150 miles of pressurized pipelines (see Figure 2-6) transporting 66 Mtpa of dense-phase CO_2 captured from industrial and natural sources, mostly for EOR, which is not a sustainable use of CO_2.[1] The Permian Basin has the most extensive pipeline network, owned by Kinder Morgan and Occidental Petroleum, carrying 38 Mtpa of CO_2 to eastern New Mexico, west Texas, and southeastern Utah for EOR (NPC 2021). Although some of these existing CO_2 pipelines have additional capacity to transport CO_2, they are not located near the most likely potential CCUS industrial clusters and CO_2 emitters, and the additional capacity is too small on its own to fulfill climate objectives.

As also introduced in Chapter 2, the United States enjoys an extensive network of freight highways, rail lines, and barge waterways that are already used for long-distance transport of bulk commodities and energy products (see Figures 2-7 and 2-8). Shipping can be more cost-effective over long distances and may be viable in the long term for transporting CO_2 from regions of the United States not located near geologic storage sites, such as the Northeast and Northwest (Jenkins 2022). Today, merchant liquid CO_2 is cost-effectively moved via trucks and trailers up to 200 miles for use in the food and beverage and other industries, although this cost is higher than CO_2 moved by pipeline. Given the long lead time needed to build dedicated CO_2 pipelines, such existing modes may be used as interim measures to transport CO_2 for utilization wherever feasible.

Many analyses indicate that orders of magnitude increase in the existing U.S. CO_2 transport infrastructure capacity, primarily to transport captured CO_2 for long-term geologic storage,[2] will be needed to make a significant contribution toward a net-zero emissions target using CCUS (see, e.g., Abramson et al. 2020; Larson et al. 2021; NETL 2015). The following sections outline considerations for incorporating CO_2 utilization opportunities in the planning and development of future CO_2 transportation infrastructure, such as the location of CO_2 sources, scenarios for multimodal transport, the hazards and safety aspects of CO_2 pipelines, and the feasibility of repurposing existing infrastructure.

---

[1] EOR does result in some long-term storage of CO_2 underground, the exact amount of which (as a percentage of CO_2 injected) depends on the size and the breakthrough, permeability, and porosity of the geology of the reservoir.

[2] Ongoing work, such as the National Energy Technology Laboratory's Carbon Storage Program (NETL 2022), performs monitoring and verification to ensure long-term CO_2 storage.

4.3.1 CO$_2$ Transport in Industrial Clusters

When designing a transportation network for CO$_2$, it is important to consider proximity to current and future CO$_2$-emission-intensive sources in the United States. Clustering CCUS infrastructure around energy supplies, power generation facilities, or ports (see Figures 2-1 and 2-8) provides opportunities to deploy projects in a manner that may be less capital intensive and less likely to have a negative impact on other resources. Thus, to take advantage of economies of scale, it could be more cost-effective to capture CO$_2$ in CCUS industrial clusters sharing common transport, storage, and utilization infrastructures. In Europe, several of such industrial clusters emitting more than 300 Mtpa CO$_2$ are either being considered or under development (ZEP 2020), and the United Kingdom's industrial decarbonization strategy prioritizes industrial cluster formation (HM Government 2021). The Infrastructure Investment and Jobs Act (IIJA, Public Law 117-58) authorizes and appropriates $3.5 billion for DAC hubs ($7 billion including nonfederal cost share), which plans for co-location of DAC projects, CO$_2$ utilization offtakers, sequestration infrastructure, and transportation infrastructure to connect them. This has been estimated by the Rapid Energy Policy Evaluation and Analysis Toolkit (REPEAT) project to result in 4.9 million metric tonnes (MMT) CO$_2$ captured per year by 2026 (REPEAT 2022). Considering the current locations of industrial emissions, point-source fossil-fuel use, and hydrogen and ammonia production, a recent analysis by the Great Plains Institute identified prime candidates for carbon and hydrogen industrial clusters (Abramson et al. 2022) (see Figure 4-2). Distributed across the contiguous United States, collectively these proposed clusters emit around 1.7 billion tonnes of CO$_2$e (gigatonnes of carbon dioxide equivalents) per annum, accounting for nearly 70 percent of the total CO$_2$ emissions from industrial and power sources (Abramson et al. 2022). For many of these proposed industrial clusters, as well as the CCS projects currently under development (see Figure 4-1), it may be beneficial to divert some of the CO$_2$ stream for utilization facilities as opposed to transporting it all to long-term geological storage, due to economics, public acceptance, or for learning through piloting and increasing scale of

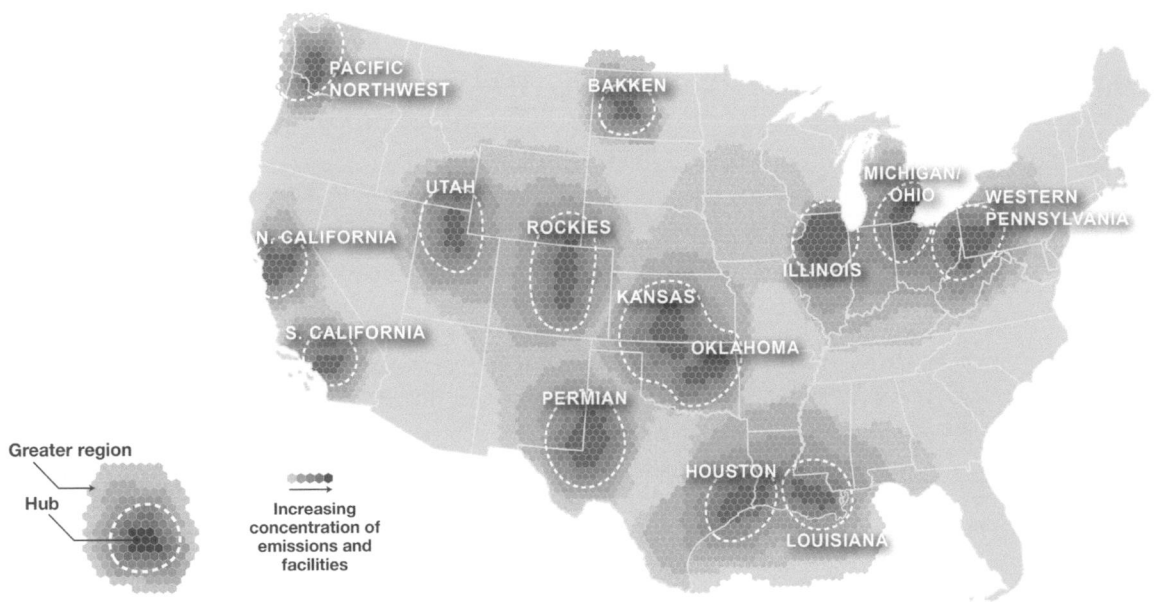

FIGURE 4-2 Map of proposed carbon and hydrogen industrial clusters in the contiguous United States, where the gray shading indicates the relative concentration of emissions and facilities.
SOURCE: E. Abramson, E. Thomley, and D. McFarlane. 2022, *An Atlas of Carbon and Hydrogen Hubs for United States Decarbonization, Minneapolis, MN: Great Plains Institute,* https://scripts.betterenergy.org/CarbonCaptureReady/GPI_Carbon_and_Hydrogen_Hubs_Atlas.pdf.

utilization. However, the feasibility of doing so will depend on the composition of the captured CO_2, purification costs, market demand, and policy incentives. Additionally, as discussed further in Chapter 6, some of the CO_2 sources in industrial clusters are not sustainable for pairing with utilization of short-lived products and will need to be replaced with DAC or biogenic CO_2 in the long term as displacement of emissions yields decreasing returns and as fossil point sources are phased out over time.

4.3.2 Multimodal CO_2 Transport Infrastructure

Besides transporting CO_2 by pipeline, CO_2 also can be transported in trucks, trains, and ships. Table 4-7 presents an overview of condition requirements by mode of transport (Al Baroudi et al. 2021). Tanker trucks are used primarily for bulk transportation of CO_2 for retail purposes. Such motor carriers store liquid CO_2 in cryogenic vessels typically at 17 bar (246 psi) and –30°C (Deng et al. 2019) and have storage capacities of 20 to 30 tonnes. Tanker trucks are considered a flexible modality, which may be relevant when small amounts of CO_2 are needed. Railroad systems have a larger transport capacity. Literature on the use of rail is limited compared to other types of CO_2 transport, but it appears more competitive than truck transport when the necessary infrastructure is already available. One of the few available literature studies, Stolaroff et al. (2021), estimates that for smaller CO_2 flows (<0.3 MMT per year) and for distances between 40 and 200 km (~25–125 miles), CO_2 transport by rail, if available, is preferable to transport by trucks (see Figure 4-3). However, the authors noted that rail access can be limited, and building new rail can cost $0.6M–1.2M per km ($0.96M–$1.92M per mile), which is somewhat more expensive than CO_2 pipelines. The authors also indicated that for distances below 40 km, truck appears to be the most cost-effective transport mode, while for CO_2 flow rates above 1 MMT per year, CO_2 pipelines are the best option.

CO_2 also can be transported by ship, which is most cost-effective for large distances. Ship transport likely would be in the liquid phase, as transport in gaseous and solid form is considered uneconomical. In the former case, this is due to the low density of gaseous CO_2, while in the latter it is due to the significant effort involved in loading and unloading solid CO_2 (Geske et al. 2015). CO_2 transport by ship exists today to transport food-grade

TABLE 4-7 Overview of Conditions for CO_2 Transportation Alternatives in the Literature

Transportation Method	Range of Conditions in Literature	Transport Phase	Remarks
Pipelines	48–200 bar (~695–2,900 psi) 9.85°C–34°C (49.73°F–93.2°F)	Vapor, dense, liquid, and supercritical phases	• Higher capital, lower operating costs • Low-pressure pipeline systems are more expensive than dense-phase transmission • Well established for enhanced oil recovery
Ships	6.5–45 bar (~95–650 psi) –52.2°C–9.85°C (–62.5°F–49.73°F)	Liquid	• Higher operating costs, lower capital costs • Currently applied in food and brewery industry for smaller quantities and different conditions • Enhanced sink-source matching
Trucks	17–20 bar (~246–290 psi) –30.2°C–20.15°C (–22.36°F to –4.27°F)	Liquid	• Capacity of 2–30 tonnes per batch • Not economical for large-scale CCUS projects • Boil-off gas emits 10% of the load
Rail	6.5–26 bar (~95–380 psi) –50.15°C to –20.15°C (–58.27°F to –4.27°F)	Liquid	• No large-scale systems in place • Loading/unloading and storage infrastructure required • Feasibility depends on existing rail line • More advantageous over medium and long distances

SOURCE: Adapted from H. Al Baroudi, A. Awoyomi, K. Patchigolla, K. Jonnalagadda, and E.J. Anthony, 2021, "A Review of Large-Scale CO_2 Shipping and Marine Emissions Management for Carbon Capture, Utilisation and Storage," *Applied Energy* 287(April):116510, https://doi.org/10.1016/j.apenergy.2021.116510.

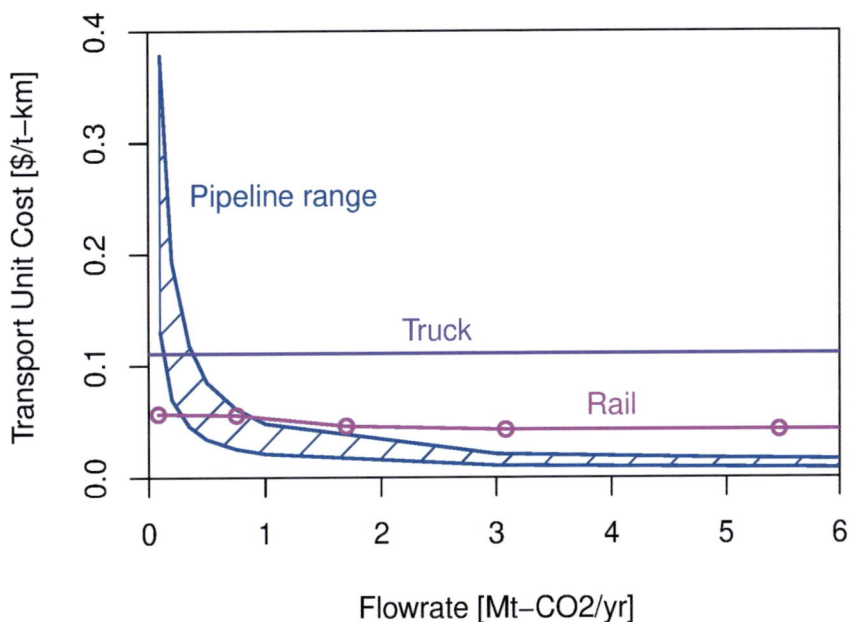

FIGURE 4-3 Comparison of costs of CO_2 transport by pipeline (blue), truck (purple), and rail (pink) for a distance of 200 km.
SOURCE: J.K. Stolaroff, S.H. Pang, W. Li, et al. 2021, "Transport Cost for Carbon Removal Projects with Biomass and CO_2 Storage," *Frontiers in Energy Research* 9(May), https://www.frontiersin.org/article/10.3389/fenrg.2021.639943. CC BY 4.0.

CO_2. Such ships typically have a capacity of about 1,000–1,500 m³ with operating pressures between 14 and 20 bar (~200–290 psi) (Jackson and Brodal 2019; ZEP 2011). Ship transportation conditions for CO_2 represent a trade-off between temperature and pressure. Roussanaly et al. (2021), for instance, pointed out that for transporting up to 20 MMT CO_2/year up to 2,000 km (1,242 miles), shipping CO_2 at 7-bar (101-psi) pressure reduces shipping costs up to 30 percent compared to shipping at 15 bar (218 psi). Bjerketvedt et al. (2022) found similar results and further reported a techno-economic limit to transport CO_2 at pressures of 15 bar beyond 10,000 tonnes due to high weight and poor capacity utilization. However, although transporting at lower pressure (7 bar, −52°C) is often recommended in the literature, to-date no commercial ship operations exist for CO_2 transport at these conditions.

Current CO_2 transport by ship occurs primarily in the food and beverage industry with ship capacities of 2,000 tonnes or less (Al Baroudi et al. 2021; Element Energy 2018; Neele et al. 2017; Reyes-Lúa et al. 2021). Given the large amounts of CO_2 expected to require transport, most studies consider that new ships with high capacities (at least 10,000 tonnes, with some studies proposing up to 100,000 tonnes) will need to be designed and built (Al Baroudi et al. 2021; IEAGHG 2004; Neele et al. 2017). It is expected that such ships can be built with design elements similar to those of liquefied petroleum gas (LPG) ships. Other options are repurposing LPG ships for CO_2 shipping or building ships in such a way that they could operate as multigas ships, for example, for transporting CO_2 as well as LPG. Such multipurpose vessels could improve business cases; however, they increase the complexity of the supply chain (e.g., vessel cleaning will be required before the ship is loaded with a different type of product) (Zahid et al. 2015). The technical and economic feasibility of repurposing existing LPG ships or of multigas operation has not yet been proven (Element Energy 2018). Finally, existing ports may be cost-effective for CO_2 shipping projects due to the availability of services and infrastructure, for example, for docking as well as storing and loading CO_2. However, the ports may impose constraints on CO_2 shipping projects, including ship length, ship draft, berth availability, and storage space requirements, that need to be considered when designing value chains (Element Energy 2018).

Transport of CO_2 in the liquid phase, regardless of transport mode, will require CO_2 liquefaction and temporary CO_2 storage facilities. A liquefaction plant is considered the most energy-consuming section in the CO_2 transport chain, demanding between 95 and 105 kWh/tonne CO_2 (Element Energy 2018; Jackson and Brodal 2019). Current industrial practices liquefy CO_2 to temperatures of −30°C and pressures of 17 bar (246 psi), generally using an external refrigerant process because of the lower unit operating energy cost. However, if CO_2 is required at different specifications, such as −52°C and 7 bar (101 psi) for large-scale commercial ship transport, much lower temperatures will be required to achieve CO_2 liquefaction than the currently employed industrial process, and other routes for liquefaction likely will be needed (e.g., high compression with free liquid expansion) (Element Energy 2018). Bjerketvedt et al. (2022) estimate that the cost of retrofitting a 15-bar (218-psi) liquefaction plant to a 7-bar (101-psi) one comprises 40 percent of the investment cost. The impact of impurities in the liquefaction process also requires attention. Deng et al. (2019) indicated that the need to purge impurities can increase the liquefaction cost by up to 34 percent compared to pure CO_2, which can influence the selection of an optimal transport mode, the location of the CO_2 utilization facility, and the economic viability of the product. An intriguing option to consider is repurposing current liquefied natural gas (LNG) facilities for CO_2 liquefaction, given the existing and increasing capacity of natural gas liquefaction capacity in the United States. U.S. LNG export capacity has increased significantly in recent years, from about 1 billion cubic feet per day in 2015 to nearly 11 billion cubic feet per day by the end of 2021 (EIA 2022b). Concurrently, liquefaction capacities for LNG have been increasing, with new facilities coming online and under construction (EIA 2022c), and are expected to continue increasing from 69.9 Mtpa in 2020 to 205.1 Mtpa in 2024 (Gulf Publishing Holdings LLC 2022). However, the committee found no information on repurposing LNG facilities for CO_2 liquefaction in the literature, and therefore could not assess the feasibility of this option, although an important consideration will be the much higher liquefaction temperature of CO_2 (−52°C) as compared to LNG (−162°C).

Another common element among transport chains for different CO_2 transportation modes is the need for temporary CO_2 storage. Liquid CO_2 can be stored in a cryogenic tank for long durations, as long as heat leaks into the tank are minimized (Lee et al. 2012). The size of the temporary storage will depend on the type and requirements of transport mode. For shipping, the capacity of CO_2 temporary storage in the port has been reported to be between 100 and 150 percent of the ship size (ZEP 2011). The transport mode dictates the pressures required for cryogenic storage. For shipping, temporary storage could be at 7 or 15 bar (101 or 218 psi) depending on the shipping conditions, as noted above. Bjerketvedt et al. (2022) indicate that the capital expenditure of buffer tanks at 7 bar is significantly lower than at 15 bar (€478 versus €867 per tonne CO_2, respectively) due to the thicker tank walls required at higher pressures. As CCS and CCU develop, the need for temporary storage will move from "simple" storage tanks toward CO_2 terminals with large-volume influxes. The extensive industry experience operating and managing LNG terminals may be used as guidance for the development of CO_2 terminals (Element Energy 2018; Myers et al. 2012; Zahid et al. 2015).

As mentioned above, in many instances, a multimodal CO_2 transport infrastructure involving ships, trucks, trains, and pipelines probably will be required. Multimodal transport is common in the industrial gas industry today, with the gas product typically railed from production to a rail depot and trucked to the last mile. Most of the industrial gas industry uses a production allocation modeling system to optimize distribution modes. Distance from production site to customer is the primary factor considered, but other aspects taken into account include feedstock costs, taxes and other financial factors, the capacity of the plant, and scheduling. Multimodal transport will especially be needed for small-scale emitters in proximity of hubs or utilization sites to enable CCU and for those located far from main pipelines where the construction of dedicated CO_2 transport and conversion plants will not be economically viable. In such circumstances, the optimal transport solution can be determined by identifying key indicators associated with cost, flexible operation, safety, and environmental impact, using multicriteria optimization models. Such an approach will involve cost-benefit analysis of a CO_2 transport network, utilizing the different modes for CO_2 transport, to minimize costs and environmental impact, reduce risks of failure, and maximize the global CO_2 utilization potential in the region. Life cycle assessment of CO_2 transport infrastructure in the selected geographic areas may also incorporate safety and environmental impacts.

Optimizing road transport of CO_2 requires solving a vehicle routing problem to determine the size of a fleet of vehicles needed to transport CO_2 to interim storage hubs or utilization plants. The design of a pipeline network operating at different pressure levels, alongside other transport modes, represents a significantly more sophisticated transportation system than has been conceived thus far in the deployment of CCUS infrastructure. Understanding the interaction between the different modes of CO_2 transport is important, as their major cost components differ: for pipelines, capital cost is the main driver, whereas for ships, trucks, and trains, operating costs make up the bulk of total cost. Reducing the purity of CO_2 streams admitted into various parts of a multitiered transportation network may also help reduce costs of CO_2 transportation. It is therefore essential for the transport system designers to understand and formulate how the synergies between the different modes of transportation of CO_2 can be used to drive down costs, while taking into account safety and minimizing carbon footprint, in order to accelerate the rollout of CO_2 utilization through the development of its supply chain. Low-carbon multimodal transportation path optimization methodologies developed for cargo transport (see, e.g., Zhang et al. 2021) can be a useful starting point for developing similar methodologies for CO_2.

4.3.3 CO_2 Transport Safety Considerations

CO_2 is a colorless, odorless gas that is heavier than air. The legal airborne concentration limit is 5,000 ppm averaged over an 8-hour shift; concentrations of 10 percent (100,000 ppm) or more can produce unconsciousness or death due to asphyxiation (EPA 2000; New Jersey Department of Health 2016). CO_2 can collect in depressions in the land, in basements, and in other low-lying areas such as valleys near a pipeline route, presenting a significant hazard if leaks continue undetected. Because CO_2 is inert, it can remain undetected for a very long time. The captured CO_2 also may contain, albeit in small concentrations, potentially toxic substances such as H_2S, whose natural dispersion might be impeded by the dense CO_2 vapor layer close to the ground, further increasing hazards (see, e.g., Soraghan and Anchondo 2022). Furthermore, upon rapid expansion, CO_2 can cool down to temperatures as low as −70°C, resulting in frostbite upon exposure, or may lead to the much more serious propagating brittle pipeline fracture for as long as the steel temperature remains below its ductile-to-brittle transition temperature (Talemi et al. 2016). Technical approaches to dealing with these safety risks are described below, but equally important in pipeline, as in other infrastructure, development is to consider the environmental, justice, and societal acceptance aspects of safety, which are discussed further in Chapter 5.

Given the above, the risks and consequences associated with the accidental release of CO_2 during its transportation must be determined. Trucking tankers are routinely monitored and checked for CO_2 leaks. Railcars often are used for transport of hazardous chemicals, so rail crews have general HAZMAT training; however, the training is not CO_2-specific, and railcars are frequently stored in railyards without attendants.[3] Of the various transportation modes, long-distance, pressurized pipelines pose the greatest hazard in the unlikely event of failure, given the large amounts of CO_2 being transported[4] and the potential for leaks remaining undetected for hours to days. These risks are especially notable for pipelines that pass near populated areas. The U.S. Pipeline and Hazardous Materials Safety Administration began regulating supercritical CO_2 pipelines in 1991, mostly for EOR applications, although it has the authority to order corrective action to address an identified safety issue on both supercritical and gaseous CO_2 pipelines. Notably, the majority of U.S. CO_2 pipelines traverse only low population areas. Despite their excellent safety record, it is not possible to draw a meaningful statistical representation of their associated risk due to the small number of CO_2 pipelines currently in operation.

A considerable body of work and understanding exists regarding the hazards associated with the high-pressure transportation of CO_2. Examples of relevant work include the European Commission FP7 project, CO_2PipeHaz (Woolley et al. 2014), which involved medium-scale CO_2 pipeline rupture tests and led to the development and validation of atmospheric dispersion models for calculating minimum safe distances to pressurized pipelines. Another important output of this work was the publication of Good Practice Guidelines for CO_2 pipeline safety (Wilday et al. 2014). In a separate study, using dispersion and risk contours mapping, Knoope et al. (2014) showed that dense-phase CO_2 pipeline transport leads to shorter lethality distances and smaller locational risks than gaseous CO_2 pipeline transport. These differences are attributed to the large momentum behind a dense-phase CO_2 release, which leads to a smaller but higher jet and a higher mixing rate with the surrounding air than a gaseous CO_2 release. Knoope et al. (2014) note that locational risks for gaseous CO_2 pipeline transport can be significantly

---

[3] Aaron Sandquist, Linde CO_2 product manager, personal communication with committee.
[4] For example, over 40 Mt for a typical 100 km (62 mile) long, 0.8-m-inner-diameter pipeline transporting CO_2 at 70 bar (1,015 psi) at 20°C.

reduced if risk mitigation measures are applied, such as burying the pipeline at 2.0-m depth, installing marker tape, and increasing surveillance.

The eventual transformation of a small puncture into a much more serious propagating brittle facture given the very low temperatures reached during the slow depressurization process was recently identified as being likely for CO_2 transportation pipelines (Martynov et al. 2017). However, field tests of sufficiently large scale and duration to lead to such type of failure in practice have not been conducted. The required tests involve the construction of large-scale pipeline test facilities (of the type recently constructed in China, funded by the European Commission and industry; see CO_2Quest 2017), which are capital intensive and require industry participation and commitment.

Of the typical impurities found in the captured CO_2 stream, water and H_2S are particularly problematic for presenting hazards during pipeline transportation. Water presents the risk of corrosion due to the formation of carbonic acid, as mentioned previously, and H_2S presents the risk of stress corrosion cracking on its own or with corrosion in the presence of water (Zeng and Li 2020). The presence of oxygen (O_2) can act as a promoter of corrosion. Experiments have shown that the presence of both sulfur dioxide (SO_2) and O_2 significantly increases the corrosion rate of carbon steel under CO_2 supercritical conditions (Choi et al. 2010). Chemicals in the vapor state at typical pipeline operation conditions are referred to as noncondensable gases (NCGs). Potential NCGs in the case of CCS are nitrogen (N_2), hydrogen (H_2), argon (Ar), O_2, and methane (CH_4). Ar, N_2, and O_2 are likely to be present in CO_2 flows originating from oxyfuel plants, while flows from pre-combustion capture are likely to contain H_2 and CH_4. The introduction of NCGs can affect the phase behavior of the mixture. Inert gases such as N_2, H_2, and Ar impact the CO_2 flow properties and influence the amount of CO_2 that can be transported in two ways: by taking up volume and influencing the flow velocity. Moreover, inert gases may increase the amount of energy required for compression, as their compressibility factors are higher than that of CO_2. NCGs also increase the saturation pressure of CO_2, thereby increasing the risk of propagating ductile fractures (Mahgerefteh et al. 2012).

Impurities in the CO_2 stream also present the risk of transition into two-phase flow, which must be avoided for a number of reasons. First, it may cause cavitation in the pipeline, resulting in the momentary opening of in-line non-return isolation valves or causing compressor trips (Peletiri et al. 2019). Second, when dense CO_2 turns into gas, the transition results in a rise in pressure, which may be above the pipeline's safe design pressure. Furthermore, the density of gaseous CO_2 is much lower than that of dense phase CO_2, which significantly reduces the transport capacity.

4.3.4 Repurposing Natural Gas Pipelines for Transporting CO_2: Opportunities and Challenges

In addition to the potential construction of new CO_2 pipelines, repurposing existing hydrocarbon pipelines to transport CO_2 is being considered as a possible means of reducing surface impacts and capital expenditure costs (Nickel et al. 2022). Notwithstanding overcoming the regulatory issues associated with the different hazard profile of CO_2 compared to hydrocarbon pipelines, repurposing the existing hydrocarbon pipeline assets could reduce implementation time by simplifying acquisition of rights-of-way, easements, or other necessary landowner agreements. With over 495,000 miles of pressurized hydrocarbon pipelines, the United States is in a unique position to consider taking advantage of such resources for transporting CO_2. Figure 4-4 shows the locations of natural gas and CO_2 pipelines in the United States, with the majority (>300,000 miles) being for natural gas. The transition away from natural gas to meet the 2050 net-zero-emissions target will mean that many of these valuable pipeline and distribution assets may become available, and potentially could be used for transporting CO_2; however, coordinating the shift from natural gas to CO_2 could be logistically challenging.

Evaluating the feasibility of converting hydrocarbon pipelines to transport CO_2 will require addressing a number of pipeline integrity and operational issues. As mentioned earlier, the most efficient way to transport CO_2 by pipeline is either in its dense phase (below its critical temperature of 31.04°C and above its critical pressure of 1,070 psi) or in its supercritical phase (above both the critical temperature and pressure) (Wang et al. 2019). In both of these states, CO_2 advantageously exhibits both liquid-like (high-density) and gas-like (low-viscosity) behavior. However, when possible, transporting CO_2 in the dense phase as opposed to the supercritical state is advantageous, given the higher fracture toughness requirement in the case of the latter (Talemi et al. 2019).

Most natural gas pipelines are rated below 1,070 psi, so the CO_2 would need to be transported in the gaseous

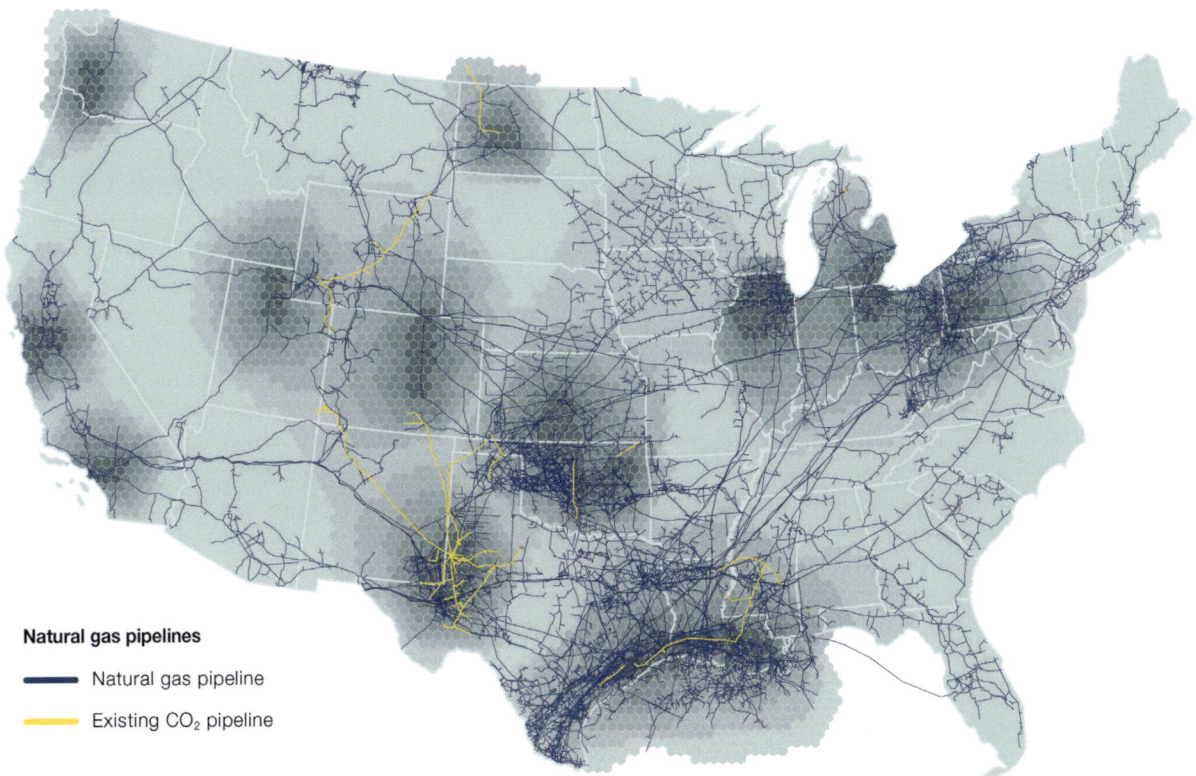

FIGURE 4-4 Map of natural gas (blue) and CO_2 (yellow) pipeline networks in the contiguous United States.
SOURCE: E. Abramson, E. Thomley, and D. McFarlane. 2022, *An Atlas of Carbon and Hydrogen Hubs for United States Decarbonization,* Minneapolis, MN: Great Plains Institute, https://scripts.betterenergy.org/CarbonCaptureReady/GPI_Carbon_and_Hydrogen_Hubs_Atlas.pdf.

state, thereby increasing the compression power requirements. (Retrofitting natural gas pipelines to increase the pressure rating is not possible.) Fluid flow simulations have shown that to transport significant volumes of gas-phase CO_2 for distances over 140 km (87 miles), a large number of in-line compressors located at relatively short distances would be required, significantly increasing costs (NPC 2021). Where possible, this may be counteracted by using larger-diameter natural gas pipelines to reduce frictional losses, or a favorable elevation profile. Secondly, the presence of even small amounts of water (>500 ppm) may result in pipeline corrosion due to the formation of carbonic acid, as mentioned earlier (Brownsort 2019). As such, additional dehydration may be required for gaseous- versus dense- or supercritical-phase transport, thereby increasing costs. CO_2 has an unusually high saturation pressure; consequently, CO_2 pipelines are more prone to propagating ductile fractures than natural gas pipelines (Mahgerefteh et al. 2012). Other than increasing the pipeline fracture toughness, which is impractical, in-line crack arrestors may be required to reduce the risk of propagating ductile fractures. Furthermore, supercritical CO_2 is an excellent solvent, so the pipeline gaskets and elastomeric seals would have to be changed to new materials to reduce the risk of leaks and explosive decompression damage (Chemical Engineering 2010; Ho 2006). Notwithstanding all the issues above, in some cases, the cost to repurpose a hydrocarbon pipeline for CO_2 transportation may be significantly less than the cost of building new pipelines. For example, reuse of the Atlantic pipeline in the Acorn CCS project is estimated at €38 million, whereas the cost of building a new pipeline could be over €110 million (Alcalde et al. 2019).

4.4 CO_2 CONVERSION AND PRODUCT TRANSPORTATION

4.4.1 Impact of Feedstock and Product Properties on Infrastructure Decisions

Where possible, co-located CO_2 capture and use is the most sensible approach for CO_2 utilization to avoid the expense of CO_2 transport. CO_2 from DAC, biogenic sources (e.g., ethanol plants), and the chemical process emissions from cement plants (from calcium carbonate decomposition) are examples of sustainable sources of CO_2. Each class of CO_2 source has its own profile for co-location with utilization. DAC facilities require substantial low-cost land area and would be suitable for co-locating a CO_2 conversion unit. Hydrogen needed for hydrocarbon CO_2-derived products may be co-generated from renewable wind and solar or other zero-carbon electricity (e.g., by integrating wind farms or solar with DAC), and the resulting products transported from the facility via trucks, rail, or new or existing[5] pipelines.

Biogenic CO_2 from corn ethanol plants is produced in small quantities per plant relative to those needed for large, capital-intensive processes, for example, hydrocarbon product synthesis. Distributed ethanol plants may be good candidates for collecting CO_2 to a central site for utilization. For example, three CO_2 trunk lines have been proposed in Iowa and surrounding states to collect biogenic CO_2 sources for long-term geologic storage (Eller 2022). Despite potential challenges with intermittency in CO_2 supply, the lower cost of highly concentrated bioethanol CO_2 ($15–$25/tonne) compared to that of more dilute streams from power generation or iron and steel production ($40–$120/tonne; see IEA 2021) makes its use for hydrocarbon synthesis a potentially profitable opportunity.

Biological processing of CO_2 is well suited for a distributed, co-located scale, and in some cases can link directly to a flue gas site or DAC system (Holmgren 2022). Photosynthetic methods for biological CO_2 utilization typically are performed either in open-pond or closed-photobioreactor designs, the choice of which has implications for water and land use, as well as capital and operational costs (NASEM 2019). Other important factors to consider in infrastructure design include the requirements for nutrients to cultivate algae or other microbes, regulations on use of genetically modified organisms, and methods to increase CO_2 solubility (NASEM 2019).

Cement plants are smaller scale than refineries or petrochemical plants and inherently distributed (Cemnet 2022) due to the cost of transporting limestone feedstock. As noted in the 2019 National Academies' report *Gaseous Carbon Waste Streams Utilization*, "if mineral carbonation is to be used to produce construction materials that can replace existing materials, (1) CO_2 will need to be consumed in many discrete locations to produce a range of materials and products, and (2) the logistics of CO_2 access and transportation of manufactured products will be critical for economic viability" (NASEM 2019, p. 49). For example, the mining of materials may not take place at the cement production site where CO_2 is available for capture, and the final product may need to be delivered by truck or rail to its point of use, which could require hydrogen or clean electricity to reduce transportation CO_2 emissions. Given the distributed nature of cement manufacturing (dependent on limestone ore location), and the likely persistent use of cement and concrete in the future economy, co-location of distributed plants for cement manufacture and utilization of the CO_2 emitted from these plants is likely a durable investment. The CO_2 released from the calcination reaction of limestone is not considered fossil-fuel derived and might be accepted as unavoidable CO_2 emissions, and treated similarly to DAC or biogenic CO_2 for use in Track 2 nondurable end products. On the other hand, the CO_2 emitted from the combustion of fossil fuels to make cement is avoidable via replacement of fuel with clean energy. Alternatively, that CO_2 could be captured for incorporation during cement curing, but this would require co-location of end use with capture site, which would be challenging given the highly distributed nature of end uses for concrete.

To convert CO_2 to hydrocarbon products, the substantial amount of hydrogen required will need to be generated at or transported to the site of utilization, which can be accomplished in one of the following ways: (1) via pipeline, truck, or rail; (2) by onsite generation via water electrolysis (requiring water and electricity); or (3) formed from natural gas (sourced via pipeline) with offtake of CO_2 to storage (onsite or via pipeline), or (4) via natural gas pyrolysis to form hydrogen with solid carbon co-product for storage or use. Figure 3-8 details these options in terms of maximum power required if hydrogen is supplied by electrolysis, or maximum natural gas where both

---

[5] Existing pipelines can be used if the product is a synthetic fuel or other chemical currently transported by pipeline and if the siting is favorably located.

hydrogen and electricity are supplied from natural gas with CCS. Option 3, combining hydrogen generated by natural gas with CCS with sustainable CO_2 (i.e., from DAC, direct ocean capture [DOC], or a biogenic source), is counterintuitive from an energy and efficiency perspective but relevant to consider given current policy incentives (e.g., the 45Q tax credit, which is discussed further in Chapter 5).

4.4.2 Considerations for Transporting Supercritical or Dense Phase CO_2 Feedstock Versus a Solid or Liquid Product

When co-location of the CO_2 feedstock and utilization plant is not feasible, customer needs and the economics of the market drive decisions of where to locate facilities. Important considerations include the capacity factor of the plant, the ability to obtain economies of scale, and process flexibility to adapt to fluctuations in supply. The industrial gas industry has traditionally employed centralized hub-and-spoke models to make use of economies of scale and to drive down costs. However, future CO_2 markets may have to consider the interdependence of the CO_2 capture process and the CO_2 utilization plant. As an example, if the CO_2 utilization goes down for any reason, captured CO_2 may be vented to the atmosphere, unless buffering storage has been built into the design. Similarly, interruptions to the CO_2 capture process can negatively impact the reliability of feedstock supply to the CO_2 utilization plant. This interdependence stacks operational risks and can pose a challenge for investors and financiers. Also, different CO_2 sources may have different sustainability profiles, impacting the life cycle assessment of a utilization process and complicating source switching.

Continuous industrial processes, such as production of liquid synthetic fuels or chemicals, operate best in a steady state. Large-scale production of liquid chemicals is typically served by direct pipeline connection to the feedstock, to reduce both supply-chain disruptions and costs. As discussed in Section 4.3, CO_2 is typically transported by pipeline in the dense or the supercritical state. However, where possible, that is, when the ambient temperature is below the critical temperature of CO_2 (31.04°C), transport of CO_2 in the dense phase is preferred, given the lower pipeline fracture toughness required. To maintain CO_2 in the dense phase during transport, the inlet pipeline pressure must be maintained above the critical pressure of CO_2. The need for booster pumps depends on pipeline length and diameter (Peletiri et al. 2018). Changes in CO_2 phase, impurities, or humidity accelerate pipeline materials corrosion, corrosion product deposition, and hydrate formation, considerations that play a significant role in pipeline development projects. Pipelines can be leveraged over long distances in CO_2 hubs, networks, and clusters, as discussed in Section 4.3.1. Although this option is considered best for reliability of supply, it is also more expensive over longer distances for single point source to sink.

For solid and liquid products that can leverage existing infrastructure through either co-generation or retrofit, siting at current facilities could be economically advantageous. Synthetic natural gas can be produced at natural gas processing facilities to make use of widespread natural gas pipelines for product transfer. Generation of ethylene or propylene or other gaseous products other than synthetic natural gas would best be located at facilities that use these chemicals as feedstock; otherwise, dedicated pipelines must be constructed, which is expensive at small scale. As mentioned earlier, CO_2 mineralization plants are better located near urban centers where there is a higher demand for construction products, as opposed to location at a large industrial CO_2 emitter. An alternative option would be to locate the CO_2 mineralization plant near the source of mined minerals to react with CO_2 and then transport the products to point-of-use, but transportation of heavy solid CO_2 mineralization products is typically more expensive than moving CO_2 over long distances. Improving the logistics for the last mile delivery can help balance supply and demand and increase profitability of solid CO_2 products. Biologically based CO_2 utilization technologies are good candidates for distributed systems that can link directly to flue gas sources without CO_2 capture, or even to DAC systems, which have more flexibility of placement. For smaller and more distributed systems, such as niche CO_2 products that have smaller capacities and higher profit margins, CO_2 can be transported cost-effectively to conversion facilities as a liquid product. Depending on their proximity to the CO_2 conversion facility and the delivery locations, redundant oil pipelines may be employed to transport liquid CO_2-derived products such as methanol. However, given the much higher flammability of such products, stricter safety and mitigation procedures will need to be put into place.

4.5 ENABLING INFRASTRUCTURE

Substantial increases in infrastructure are needed to provide the electricity, hydrogen, and water required for CO_2 utilization. The net-zero economy broadly is projected to require at minimum a 40 percent increase in demand for electricity and a much greater increase in generating capacity (up to 800 percent for some scenarios, given low stream factors for variable renewable energy) (Cole et al. 2021; Larson et al. 2021; Ruth et al. 2020). A dramatic expansion in hydrogen production—using either a portion of this electricity or natural gas with CCS—will also be required to achieve economy-wide decarbonization goals, which includes use of hydrogen for energy storage (Cole et al. 2021; Larson et al. 2021; Ruth et al. 2020). Overall water use for the energy sector will diminish as steam power cycles are replaced by renewable energy, but some of this reduction will be offset by electrolytic generation of hydrogen and increased use of steam for reforming of natural gas (with CCS). CO_2 utilization will play a part in these changing energy system demands. Deployment of enabling infrastructure for CO_2 utilization is challenged because use of existing fossil point sources of CO_2 is not sustainable (i.e., not compatible with the Paris Agreement goal of limiting global warming to below 1.5°C compared to pre-industrial levels) for all utilization processes, and DAC facilities do not yet exist. This section discusses considerations for developing the extensive infrastructure that would be needed for clean electricity, hydrogen, water, natural gas, and energy storage required to support future CO_2 utilization projects. Land-use constraints that might impact infrastructure decisions also are described, with particular emphasis on trade-offs between areal land use and energy and hydrogen requirements for different methods of hydrocarbon production.

4.5.1 Clean Electricity

As introduced in Chapter 3, CO_2 capture and utilization processes require significant amounts of clean electricity. For example, synthesis of hydrocarbon fuels from CO_2 requires about four times more energy than direct use of renewable/clean electricity for a given usage scenario. Therefore, deployment of CO_2 utilization technologies will put added demands on grid expansion beyond the already significant need to expand renewable power for direct electrification of industry, residential, and transportation sectors. Different regions of the country will have different demand curves emerging as electrification continues, for example, the "duck curve" in California, driven by daily solar generation patterns (DOE-EERE 2017) versus the "butterfly curve" in New England, driven by demand during the winter months if heating is increasingly electrified (Hewitt 2019). As more renewable electricity is connected to the grid, the intermittency and relative lack of dispatchability of these sources may create an increasing temporal mismatch between demand and supply of clean electricity across hours of the day and seasons of the year. In any case, as the U.S. power grid transitions to renewable energy, energy storage must be included to enable CO_2 capture and utilization operations. With the increased stand-alone storage and hybrid capacity entering interconnection queues across organized markets as an indication (LBNL 2022), it would seem that some of the necessary grid updates are trending in the right direction, provided interconnections actually occur. Accommodating increased demand will require grid updates (transformers, high-voltage lines including high-voltage direct current, etc.) for existing electricity infrastructure and connection of new clean electricity generation to the grid (NASEM 2021). As of the end of 2021, about 900 gigawatts (GW) of wind and solar generation were in the interconnection queue (LBNL 2022), providing a sense for how much supply is coming into the market.

As discussed in Section 2.4.1, constant operation is favorable for the many current CO_2 utilization technologies that entail multiple steps, or high-pressure or -temperature operation. If a CO_2 utilization process is developed that can be economically viable while operating intermittently, it may be able to take advantage of negatively priced carbon-free electricity at times of excess supply. Ultimately, this question will be a function of energy, capital, and other costs, and a matter for grid optimization and utilization facility electricity procurement terms.

An entity attempting to reduce or eliminate the carbon intensity of its input electricity to advance net-zero emissions CO_2 utilization may locate the production facility within an electricity grid (by service provider or region) that has a low average emissions profile (see EIA 2022a, Table 6; Van Atten et al. 2021, slide 15) or in a state that has aggressive state-level decarbonization goals (C2ES 2021). A third option is to enter into a power-purchase agreement (PPA), which allows decarbonized electricity from an identified provider to be purchased from the existing grid without significant infrastructure upgrades at the point of consumption. Alternatively, the project developer could sponsor a new decarbonized electricity project (e.g., solar or wind), but this may have significant infrastructure or land-use implications at the project location and may require distribution and/or transmission upgrades.

Finally, carbon credits generated during excess renewable electricity availability may be used to offset some or all of the carbon-emitting supply to a CO_2 utilization facility to achieve 24/7 low-carbon electricity on a "net" basis. Any project involving siting or development of new electricity infrastructure will need to incorporate community engagement and consider environmental and justice impacts, best practices for which are discussed in Chapter 5.

4.5.2 Hydrogen

Developing infrastructure for hydrogen production, distribution, storage, and use requires consideration of many factors. A primary consideration is whether, for a given application, to rely on centralized or decentralized (distributed) hydrogen production. Centralized production takes advantage of economies of scale but will likely require transportation to move hydrogen from the point of production to the point of consumption. Figure 4-5 illustrates the complexity of hydrogen value chains involving transportation by truck and pipeline or by ship. On the other hand, decentralized production takes advantage of economies of location. Both centralized and decentralized hydrogen production ideally require access to a stable, 24-hour guarantee of electricity. In the case of decentralized production, access to wholesale electricity prices might not be possible, so an appropriate tariff will need to be set up to acquire the requisite electricity at a low-enough cost. Importantly, the generation mix underlying

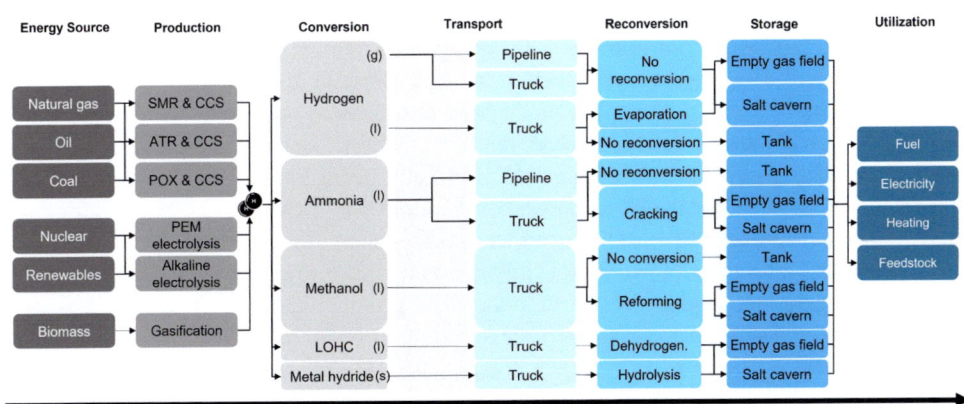

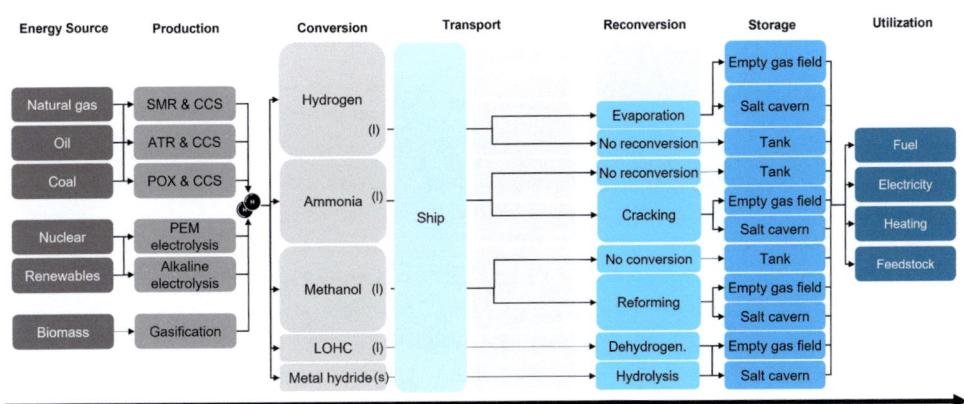

FIGURE 4-5 Components of the hydrogen value chain for various production and end uses, shown for transport by truck and pipeline (top) and for transport by shipping (bottom).
SOURCE: M. Hiestermann, 2022, "Hydrogen Supply Chain Portfolios in the Netherlands Between 2030 and 2050," Master thesis, Delft University of Technology, Netherlands, https://repository.tudelft.nl/islandora/object/uuid%3Ac4b1c997-f0f6-4859-ba9b-0f579c8d9052.

such a tariff would have to be of an emissions intensity amenable to the life cycle CO_2 requirement for the CO_2 utilization product to be low-, neutral-, or negative-carbon.

There may be opportunities to adapt existing infrastructure for hydrogen, for example, by converting natural gas pipelines for hydrogen transport or using SMR or ATR with CCS for hydrogen generation. Blending hydrogen into natural gas pipelines, at levels up to about 20 percent by volume, is being explored as an option to decrease the carbon intensity of gas-powered end uses with existing infrastructure (Blanton et al. 2021; EFI 2021; IEA 2019). However, such blends of hydrogen and natural gas are not an attractive option as a method of transporting hydrogen for CO_2 utilization, because utilization processes would require pure hydrogen. Repurposing natural gas pipelines to transport 100 percent hydrogen faces technical challenges, such as requirements for higher compression power and modifications to prevent embrittlement; case-by-case analyses are needed to determine its feasibility (ACER 2021). Clean hydrogen production from SMR or ATR with CCS in some cases can use existing infrastructure, which would allow for more rapid permitting at lower cost. These factors also provide a large incentive to co-locate CCU projects with CO_2 storage sites and their connecting pipeline infrastructure. However, hydrogen generation from SMR with CCS could result in additional local air pollution depending on the carbon capture technology used, as discussed in Section 4.1. Therefore, the impacts on the surrounding community would need to be assessed carefully and the community consulted before developing such a facility. Newer technologies for hydrogen generation from natural gas (e.g., ATR, POx) enable co-generation of clean power, such that both the process and process-energy used for hydrogen generation can be decarbonized from a single point of capture (Liu 2021).

It is also important to recognize that hydrogen has a 100-year global warming potential (GWP),[6] which is still being investigated but with estimates that range from 3.3 to 11 (Field and Derwent 2021; Warwick et al. 2022). Although hydrogen itself is not a greenhouse gas, it could alter atmospheric chemistry by reacting with hydroxyl radicals (OH•) that would otherwise react with other pollutants such as methane, thereby extending its lifetime in the atmosphere. Hydrogen's GWP can influence infrastructure decisions. For instance, infrastructure may be designed to minimize hydrogen transport, such as through distributed production, or minimize any leakage of hydrogen during transport or storage by conversion into less volatile hydrogen carriers such as methanol or ammonia. Which of these energy carriers would provide a better solution (i.e., lower GHG emissions) requires further knowledge of the total (direct and indirect) contribution of the different hydrogen carriers to the GHG potential. However, hydrogen leakage would be much smaller than the CO_2 emissions that it would replace, resulting in an overall positive climate impact. The total emissions (i.e., carbon footprint) attributed to hydrogen highly depend on the way it is produced, as shown in Mac Dowell et al. (2021) (see Figure 4-6). In the case of electrolytic hydrogen, the primary determining factor is the carbon intensity of the electricity used; if the carbon intensity of electricity exceeds 140 kg CO_2/MWhe (megawatt-hours of electricity), the hydrogen generated will have a higher carbon footprint than that produced by SMR of natural gas with CCS (Mac Dowell et al. 2021). Figure 4-6 also shows that, for hydrogen derived from natural gas with CCS, guaranteeing deployment of best practices in the natural gas supply chain is critical to achieving zero-emission targets.

4.5.3 Water

Water is a fundamental commodity for almost every step of production processes around the world. It is used as a feedstock and/or as a utility for heating and cooling products and equipment. Understanding whether a given technology's demand for water resources will compete with other water uses is fundamental in determining the impact of that technology on the full water and energy system. In the case of CCU, water is used during the capture of CO_2, to produce hydrogen, and during the production of CO_2-based products. A recent review (Rosa et al. 2021) reported the water footprint of CO_2 capture in the range of 0.5 to 3 m³ water per tonne CO_2 for point-source capture from fossil sources (for electricity generation) and 2 to 7 m³ water per tonne CO_2 for DAC with carbon sequestration. Water is used in the capture process itself, and the amount required is highly dependent on the type of capture and the cooling technology employed.

During the production of CO_2-derived products, water is used not only for cooling equipment but also as an extractive or absorptive medium or for other processes. As noted in Section 2.4.3, water availability is an

---

[6] GWP indicates the amount of energy that the emission of 1 ton of a given greenhouse gas absorbs over time (typically 100 years) compared to that for 1 ton of CO_2. As the reference gas, CO_2 is given a GWP of 1 (EPA 2022b).

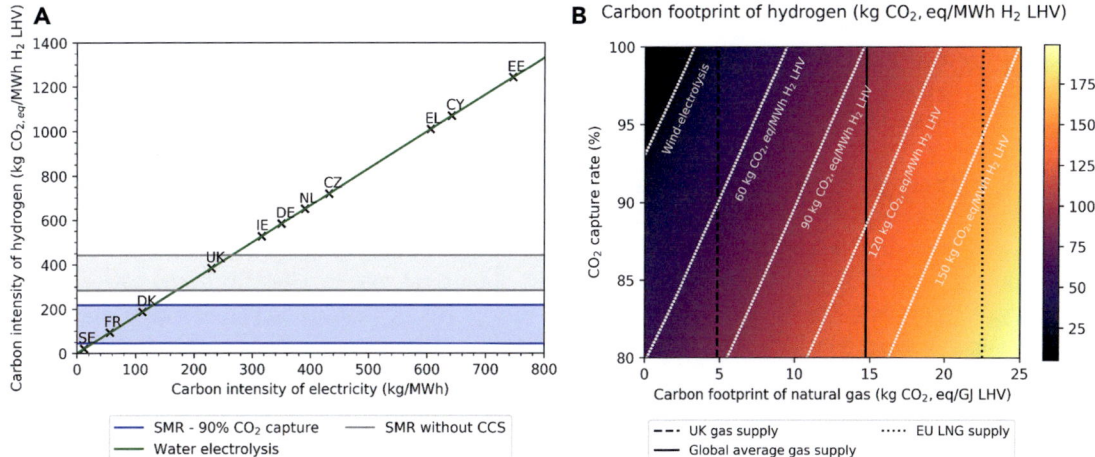

FIGURE 4-6 Carbon intensity of hydrogen produced by different methods. (Left) Comparison between methane-derived (with and without CCS) and electrolytic hydrogen, incorporating emissions from purchased natural gas and electricity. The crosses show the current average carbon intensity of electricity in several European countries. The figure indicates that for electrolytic hydrogen to have an equal or smaller carbon footprint than methane-derived hydrogen with CCS, the carbon intensity of electricity needs to be below 140 kgCO$_2$/MWhe. (For reference, as of January 2020, the average carbon intensity of electricity in the United States was 371 kg CO$_2$/MWhe; see EPA 2022a.) (Right) Comparison between electrolytic hydrogen and hydrogen derived from natural gas with CCS, and evaluation of the impact of CO$_2$ capture and natural gas supply-chain emissions, with the color scale showing the resulting carbon footprint of hydrogen. When the carbon footprint of natural gas is low and the capture rates are greater than 90 percent, the carbon footprint of methane-derived hydrogen approaches that of electrolytic hydrogen. The global carbon footprint of methane, however, results in a carbon footprint of hydrogen much larger than electrolytic hydrogen even at high capture rates.
SOURCE: Reprinted from N. Mac Dowell, N. Sunny, N. Brandon, et al., 2021, "The Hydrogen Economy: A Pragmatic Path Forward," *Joule* 5(10):2524–2529, https://doi.org/10.1016/j.joule.2021.09.014. Copyright 2021, with permission from Elsevier.

especially important consideration for biological CO$_2$ utilization processes that need water for cultivating algae and other microbes. Argonne National Laboratory, with support from DOE's Bioenergy Technology Office and Office of Fossil Energy and Carbon Management, has developed the Available Water Remaining for the United States (AWARE-US) Model (ANL 2020), which has been used to analyze the water stress impacts of siting algal cultivation systems (see, e.g., DOE-BETO 2021; Xu et al. 2019). As another example, mineralization of waste often involves high-water-containing "wet" systems, where waste particles are suspended in or dissolved in an aqueous solution. A comparison of five different mineralization processes that take place under aqueous conditions showed a large range of water requirements (mostly to form the slurry at the desired concentration), varying from 35 to 1,400 tonnes water per tonne CO$_2$ (Naraharisetti et al. 2017). Other processes that need water include vacuum production, product rinsing, dilution, and distillation. Finally, not only the amount but also the quality of the water differs per process. In processes that operate at high temperature and/or pressure, requirements are in place regarding impurities (e.g., salts) that could cause scale formation. Other processes have even stricter requirements for water quality, such as electrolytic processes for hydrogen production in proton exchange membrane (PEM) systems where high-purity water[7] is required. For PEM systems, cation impurities in the water feed are a constraint, with the most common being Fe^{3+}, Mg^{2+}, Ca^{2+}, Cu^{2+}, and Na$^+$ (Yoshimura et al. 2022).

As discussed in Chapter 3, and illustrated in Figure 3-7, many CO$_2$ utilization products, especially chemicals and fuels, will require hydrogen as a feedstock. Today, most hydrogen is produced from natural gas (through SMR), but to be compatible with net-zero-emission targets, a shift toward low-carbon routes such as water electrolysis is expected, as discussed in Section 2.4.2. Of the various methods for hydrogen production, water electrolysis

---

[7] The American Society for Testing and Materials defines Type II water, as required in commercial electrolyzers, as having a resistivity of >1 MΩ-cm, sodium and chloride content <5μg/L, and <50 ppb total organic carbon.

has the largest water intensity, as illustrated in Figure 3-8. Note that the range of values in Figure 3-8 reflects to a large extent "local" water consumption used at the production site. If all water used in upstream processes were accounted for, the values could be larger. The additional water demand throughout the supply chain is important to consider when designing large-scale infrastructure strategies. For instance, Shi et al. (2020) assessed water usage in the full supply chain of hydrogen production through electrolysis, including manufacturing of the system; production, installation, and operation of photovoltaic (PV) panels and the electrolysis system; and electrolysis operation. The authors report water consumption for green hydrogen from PV in the range of 22 to 126 kg water per kg hydrogen depending on solar radiation, lifetime, and silicon content.

At the global and national levels, there is limited concern about total water availability for hydrogen production (i.e., electrolysis operation). The International Renewable Energy Agency estimated that to supply approximately 409 million tonnes hydrogen by 2050 (the projected need in the 1.5°C pathway called for in the Paris Agreement, see UNFCCC 2022), about 7–9 billion m^3 water per year will be required, which corresponds to less than 0.25 percent of current total freshwater consumption annually worldwide (IRENA 2022). Although this indicates that, at a global level, water availability is not a showstopper for hydrogen production, the regions where low-carbon electricity can be produced most efficiently (e.g., regions with reliably high incoming solar radiation loads) also tend to be the driest. Tarroja et al. (2018) report that water availability can limit solar thermal utilization in California to between 11 and 53 percent of its solar thermal potential and conclude that the spatial distribution of water will drive the extent to which California can use its solar thermal and geothermal resources. Devitt et al. (2020) report a similar result for west Texas and point out that changes induced by climate change may lessen the potential further. A recent study in the United Kingdom estimates a 5 percent national increase in freshwater consumption due to the implementation of a hydrogen economy; however, local impacts in hydrogen hubs may be as high as 20 percent (Hopcroft and Papadamou 2021). In the United States, local water impacts of hydrogen production are just starting to be investigated (Rustagi 2022) and will need to be considered in the context of environmental justice and equitable deployment of energy infrastructure.

Furthermore, Rystad Energy (2021) reported that over 85 percent of the green hydrogen capacity in 2040, based on projects planned worldwide as of mid-2021, may need to source water via desalination, which would add to the cost of hydrogen and impose additional demand for clean electricity to keep a low carbon footprint for hydrogen. Increasing deployment of desalination facilities could, on the other hand, decrease local water stress indices by providing communities with water not only for hydrogen production but also for other uses. Additionally, it may be possible to co-locate a desalination plant with a DOC facility in order to decrease the costs and energy consumption of CO_2 capture (Digdaya et al. 2020; Eisaman et al. 2018). Finally, another possibility to mitigate local impacts of hydrogen production in freshwater availability is the development of seawater electrolysis. This option is still in its infancy, however, and will require significant R&D before commercialization (Khan et al. 2021).

Fresh or potable water availability must be considered in implementing CO_2 capture and utilization projects, including supply of electricity and hydrogen. If hydrogen is generated at the CO_2 utilization manufacturing site, then the required water must be supplied on-site. Alternatively, hydrogen may be generated at a centralized facility using the indicated amount of water and transported to the manufacturing site by hydrogen pipeline, rail, or truck. Many conventional power cycles consume water as withdrawals via steam cycles used in power generation, though direct use of solar or wind energy will reduce this need (Lampert et al. 2015, 2016). Some DAC facilities may either consume or co-produce water in a sorbent-regeneration step. The total amount of water needed for the full plant will depend on the water needs for the steam regeneration unit (when heat is required for regenerating the sorbent) and the pure water generated as by-product. Among the most water-intensive approaches are capture methods based on humidity swing absorption, for instance, which use water as the regeneration medium. Van der Giesen et al. (2017) report water requirements for this DAC system on the order of 15.2 tonnes water per tonne CO_2 capture. For other types of DAC systems, values are lower; for example, Keith et al. (2018) report water requirements of 4.7 tonnes water per tonne CO_2 captured for a DAC system based on an aqueous potassium hydroxide sorbent, while Ozkan (2021) reports 1–7 tonnes water per tonne CO_2 capture for similar systems.

4.5.4 Land-Use Constraints

Increased deployment of CO_2 utilization could have substantial land-use impacts depending on the CO_2 source and utilization process. Existing hydrocarbon production largely uses carbon and energy feedstocks derived from subsurface resources of oil and gas that are high energy density and have a low areal footprint. Sustainable CO_2-derived hydrocarbon products require CO_2 from DAC or biogenic CO_2 derived from currently limited biofuels or bio-based chemicals production. DAC requires less land area than the equivalent use of first- or second-generation bio-based feedstocks, but more land than associated with production from oil and gas.

Figure 4-7 presents the committee's analysis of relative land-use requirements for different methods of hydrocarbon production. For illustration, units of $-CH_2-$ represent the hydrocarbon product such as synthetic diesel or hydrocarbon polymers. This analysis assumed hydrogen generation via electrolysis using 50.4 MWh of electricity per ton hydrogen (Bazzanella and Ausfelder 2017), and used the estimates given in Figure 3-7 of kilograms hydrogen required to react with CO_2 to generate hydrocarbon products. For synthetic fuel or ethylene produced from CO_2 by DAC, estimated land-use requirements (per million tonnes fuel per year) range from 400 to 1,600 km² (150–600 mi²), relative to 3,500 km² (1,350 mi²) from biomass, 0.2 to 0.5 km² (0.07–0.2 mi²) from natural gas or oil, and about 30 km² (12 mi²) if the CO_2 were stored (sequestered) instead of transformed into hydrocarbons (labeled as "DAC only" in Figure 4-7). These estimates show the increase in land use associated with production of synthetic fuels or chemicals from CO_2, relative to current production from traditional petroleum or natural gas, and compared to the options of storing the captured CO_2 or producing fuels from biomass. Land use for synthetic fuels or chemicals from CO_2 is somewhat less than use of biomass. Land-use and ecosystem impacts of unconventional oil and gas production ("fracking") can be larger (Allred et al. 2015).

Land and infrastructure for DAC are largely nonexistent today, as the technology is at the early stages of demonstration and development. Land-use requirements for renewable energy are also much greater than for fossil

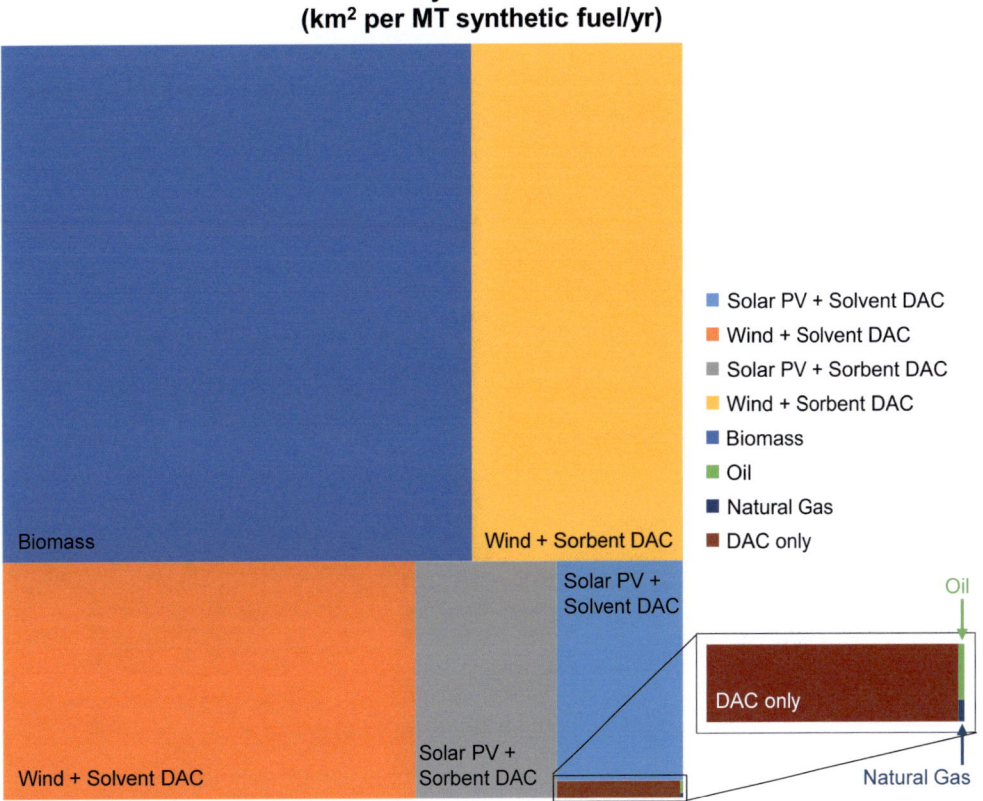

FIGURE 4-7 Estimates of land-use requirements for different methods of hydrocarbon production. Land-use requirements for producing hydrocarbons from CO_2 are larger than those for traditional production methods using oil and gas, but smaller than those for fuel production from biomass (based on median power densities for electricity generation from these sources). SOURCES: Committee generated based on data from Bazzanella and Ausfelder (2017); Lebling et al. (2022); Ozkan et al. (2022); van Zalk and Behrens (2018).

energy and chemicals, which adds to usage requirements for DAC and other renewably-powered technologies. Any CO_2 utilization process requiring hydrogen may need additional land for hydrogen production, depending on how and where that hydrogen is produced. Land-use impacts would be particularly large for hydrogen production via biomass gasification, since biomass is a low-density feedstock with significant land requirements for its growth. Land considerations might be a factor in some brownfield developments and are likely to be a significant consideration for many potential new development projects when implications and needs for conveying infrastructure, pipes, wires, rail, and road transport are considered.

To understand the scale and R&D challenges of deploying industrial-scale "Power-to-X" systems, Smith et al. (2019) estimated the energy and size requirements for a 10,000-tons-per-day methanol plant based on CO_2 from DAC and hydrogen from PV/water electrolysis. Figure 4-8 compares the relevant scale of each component of the proposed system to the largest current installed or planned facility for each technology, showing that orders of magnitude increases in solar photovoltaic land area, cross section of CO_2 direct air capture, and electrolyzer catalyst surface area would be required beyond the scale of current implementations of those technologies to achieve the desired 10,000-tons-per-day capacity. Given these results, Smith et al. (2019) suggest that decentralized, smaller plants may be more practical and economical for Power-to-X processes, though the optimal solution will also depend on resource availability, geography, and political factors.

4.5.5 Energy and Hydrogen Storage

Facilities that convert CO_2 to hydrocarbon products via traditional thermochemical routes are capital intensive and therefore must operate 24/7 to be economical. Storage of energy and hydrogen may be required to achieve constant operation. If variable renewable energy (e.g., wind and solar) is used to power these facilities or to manufacture hydrogen, then energy storage probably will be required, which will add to the cost. For larger-scale manufacturing, seasonal energy storage in the form of hydrogen stored in salt domes could enable economical hydrocarbon production, but the manufacturing facilities would need to be located at or within a transportable distance from the salt dome, which imposes geographic limitations. An example of this approach is the HyDeal project, which plans to connect green hydrogen production in the Los Angeles basin with salt dome storage in Utah via a dedicated hydrogen pipeline (Green Hydrogen Coalition n.d.).

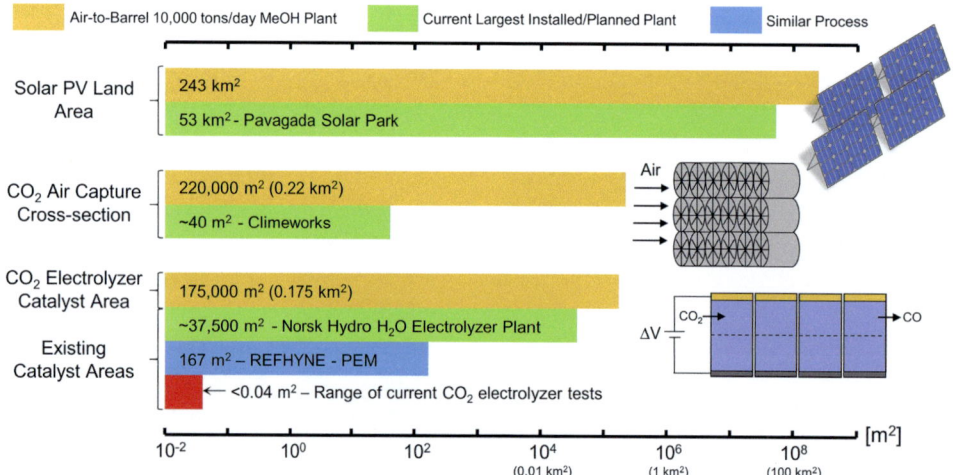

FIGURE 4-8 Scale needed for solar photovoltaic (PV) land area, direct air capture cross section, and electrolyzer catalyst surface area of a DAC-based methanol plant with 10,000-tons-per-day capacity (yellow bars) compared logarithmically to the scale of existing or planned plants and installations around the world (green, blue, and red bars).
SOURCE: Reprinted from W.A. Smith, T. Burdyny, D.A. Vermaas, and H. Geerlings, 2019, "Pathways to Industrial-Scale Fuel Out of Thin Air from CO_2 Electrolysis," *Joule* 3(8):1822–1834. Copyright 2019 with permission from Elsevier.

Electrochemical energy storage (batteries) can be deployed for short duration (4.5–13 hours) but adds cost to projects that include dedicated renewable power generation, or to power purchase agreements. Hydrogen storage via salt domes provides lower cost for longer-duration storage of renewable energy versus use of tanks, spheres, or generation of a carrier such as ammonia or methanol (Penev et al. 2019). Otherwise, it may be more cost-effective for third-party entities to store the overproduction of electricity to then be used as a buffer against real-time high marginal prices. Ultimately, this question is a function of grid optimization and utilization facility electricity procurement terms.

4.6 FINDINGS AND RECOMMENDATIONS ON INFRASTRUCTURE CONSIDERATIONS FOR CO_2 UTILIZATION

FINDING 4.1 CO_2 Capture. The cost of CO_2 capture is a deterrent to more widespread implementation of CO_2 utilization technologies. The cost of the CO_2 feedstock for utilization can vary significantly but largely depends on the concentration of CO_2 in the emission source and the type of CO_2 capture technology deployed. Today's commercial CO_2 capture facilities are deployed at high-purity CO_2 sources with mature capture technologies and produce relatively low-cost CO_2 as an output. Reducing CO_2 output costs for other capture technologies and for application at less concentrated CO_2 sources will require further development to lower capital expenses, improve technology performance, enable better integration of the capture technology with the emission source, and/or reduce energy requirements for capture.

FINDING 4.2 CO_2 Purity Requirements. CO_2 capture, transport, utilization, and storage all have different CO_2 impurity tolerances. Satisfying the purity requirements of all stakeholders across the carbon capture, utilization, and storage value chain, while also prioritizing safety and environmental considerations, is a major challenge. In the United States, purity requirements often are dictated by the midstream transportation company through a contractual obligation, though some impurities that could adversely impact CO_2 conversion are not included in such agreements.

FINDING 4.3 CO_2 Purity Needs for Utilization. CO_2 waste streams may need to be purified to meet the requirements of the CO_2 utilization technology for conversion into the desired product. The impurities present will play a key role in dictating whether and for what type of utilization the CO_2 stream is suitable. Of the various CO_2 utilization processes, mineralization and biological CO_2 conversion are the most resilient to impurities, while electrochemical and thermochemical CO_2 conversions are the most sensitive. Technologies for purifying CO_2 waste streams are commercially available but their use will increase the costs of either the CO_2 capture processes or the CO_2-based products, depending on the stage at which purification takes place.

FINDING 4.4 Liquid CO_2 Transport. The high costs and energy needs of liquefaction facilities to prepare liquid CO_2 for transport could impact site selection of a CO_2 utilization project and the optimal transport mode to access that project site.

FINDING 4.5 Repurposing Liquefied Natural Gas Facilities for CO_2. In recent years, U.S. exports of liquefied natural gas (LNG) have increased significantly, and the capacity of natural gas liquefaction facilities has been growing concurrently, with additional facilities under development. Repurposing natural gas liquefaction facilities for CO_2 liquefaction or co-designing a new facility for both natural gas and CO_2 liquefaction has not been studied extensively. If feasible, these options could prevent stranded assets if demand for natural gas decreases as the world transitions to a net-zero-emissions future.

> **RECOMMENDATION 4.1.** Agencies with authority over LNG facilities—including the U.S. Department of Energy, the Pipeline and Hazardous Materials Safety Administration, the Federal Energy Regulatory Commission, and the U.S. Coast Guard—should collaborate to assess the possibility of repurposing existing

LNG facilities for CO_2 liquefaction, as well as the feasibility of co-designing any future facilities to be able to compress both natural gas and CO_2.

FINDING 4.6 Multimodal CO_2 Transportation. Determining the optimal (potentially multimodal) CO_2 transportation method for a given CO_2 utilization project requires considering the location and type of CO_2 source, proposed sites of utilization, and estimated product volumes in order to minimize cost, environmental and justice impacts, and safety concerns. Pipeline networks can transport large volumes of CO_2 and benefit from economies of scale, but they require aggregation of CO_2 sources and may face safety and regulatory challenges if being developed near residential areas, railroads, and/or utilities. For facilities that emit smaller volumes of CO_2 or are located far from existing pipeline networks, on-site CO_2 conversion or multimodal transport of the CO_2 by truck, train, a dedicated pipeline, and/or ship likely will be more economically viable than developing a pipeline network to connect these emission sources to a remote storage or utilization site. These decisions will depend in part on how each transportation method impacts the overall CO_2 footprint of the capture and utilization process.

> **RECOMMENDATION 4.2.** The U.S. Department of Energy should develop dedicated methodologies for optimizing multimodal transport of CO_2 for utilization purposes to assist transport system designers and project developers in their planning. These methodologies should prioritize safety and environmental justice, minimize carbon footprint, and incorporate existing knowledge and lessons learned from cargo transport and other gas transportation.

FINDING 4.7 CO_2 Transport Safety. CO_2 is an asphyxiant gas but otherwise nontoxic, as well as nonflammable. There is already a considerable level of understanding of the risks associated with pipeline transportation of CO_2, including for different types of CO_2 pipelines that have different risks related to impurities. For example, the natural CO_2 sources used for enhanced oil recovery (EOR) can contain toxic hydrogen sulfide gas, whereas the CO_2 streams transported for CO_2 utilization will be higher purity. Most of the practical experience associated with CO_2 pipeline operation has been confined to EOR applications. Given the relatively small number and remote locations of these pipelines, it is difficult to draw meaningful conclusions about their failure statistics. No major safety issues are envisaged provided that appropriate mitigation steps are put in place. However, given the unique thermophysical properties of CO_2, special consideration needs to be given to selecting pipeline materials with appropriate mechanical properties to resist ductile and brittle propagating fractures, the latter of which has not received considerable attention.

> **RECOMMENDATION 4.3.** In collaboration with national laboratories and university researchers, the U.S. Department of Energy, the Pipeline and Hazardous Materials Safety Administration, and industry should co-fund research to develop rigorous fluid-structure models validated by large-scale field tests to better understand the complex processes leading to propagating brittle fractures in CO_2 pipelines, including its typical stream impurities, and provide practical solutions for avoiding them.

FINDING 4.8 Repurposing Natural Gas Pipelines for CO_2. Given the large number of parameters involved, the economic and operational feasibility of repurposing existing natural gas pipelines for transporting CO_2 requires examination on a case-by-case basis using rigorous systems analysis, with techno-economic, safety, environmental, and justice considerations being the overarching factors.

FINDING 4.9 CO_2 and Product Transportation Infrastructure. The economics of infrastructure placement will be dictated in part by the ease of transporting CO_2, hydrogen, electricity, and other inputs, versus the ease of transporting the carbon-based products. For example, aggregates are expensive to transport and so will favor distributed production, likely close to urban centers where most concrete and aggregates are used.

FINDING 4.10 CO_2 Transportation Infrastructure. The optimal CO_2 transport and delivery infrastructure to enable utilization depends on the product type. In some cases, commodity chemicals, plastics, and fuels derived from CO_2 can be synthesized using existing facilities for petrochemical-based production and would benefit from centralized infrastructure for CO_2 capture, transport, and conversion. On the other hand, concretes and aggregates are produced locally, and would be best served with a smaller-scale, local or regional CO_2 capture and/or distribution network. Clarity is needed on the long-term availability and desirability of utilizing CO_2 from point sources (fossil or biogenic) versus sourcing CO_2 from air or water, as well as the system-level trade-offs that result from developing infrastructure at different locations and scales.

FINDING 4.11 Electricity Requirements for CO_2 Utilization. Many CO_2 utilization processes will significantly increase demand for electricity, particularly zero-carbon-emissions electricity, compared to current production methods for the same product, and likely require 24/7 operation to be economical. These requirements will impact the optimal power generation mix, load management, and transmission and distribution planning for the electricity grid, especially given that many zero-carbon-emissions electricity sources provide intermittent generation and therefore likely will require concomitant development of energy storage.

> **RECOMMENDATION 4.4.** National policy should prioritize carbon-emissions-free energy as inputs to all aspects of a net-zero-carbon-emissions system, including growth in emissions-free energy to accommodate CO_2 utilization. In the near term, the U.S. Department of Energy should coordinate efforts to advance CO_2 utilization with carbon-emissions-free energy projects, especially those with intermittent characteristics, such as solar and wind energy systems that offer opportunities to capitalize on production capacity that would otherwise be curtailed.

FINDING 4.12 Hydrogen Requirements for CO_2 Utilization. Many CO_2 utilization processes require net-zero-emissions hydrogen as an input. Hydrogen is difficult to transport and store (though proven technologies exist for both), and studies on its global warming potential are evolving. These factors may impact decisions about the design of hydrogen infrastructure, favoring on-site, on-demand production of hydrogen where feasible and potentially motivating the use of hydrogen carriers (which themselves could be products of CO_2 utilization) for long-distance transportation.

> **RECOMMENDATION 4.5.** Given the complexity of transporting and storing the hydrogen required for upgrading CO_2 to hydrocarbon products, project planners should consider co-location of hydrogen generation with manufacturing plants utilizing CO_2 as a feedstock. Given that most utilization projects will require 24/7 plant operation to be economically viable, project planners should also incorporate energy storage into the facility design.

FINDING 4.13 Water Requirements for Different CO_2 Utilization Processes. In the near term, CO_2 utilization processes requiring water include electrolytic hydrogen production for reaction with CO_2, biological CO_2 conversion, and process cooling. In the longer term, (photo)electrochemical conversion of CO_2 to chemicals and fuels also may be deployed commercially and would require water as a direct input.

FINDING 4.14 Local Water Impacts of CO_2 Utilization. In total nationally, the water requirements for CO_2 utilization do not represent a significant increase above current water demand, but water availability may be limited at a local level due to geographic variability. Many regions that have the abundance of clean electricity generation capacity needed to support net-zero or net-negative CO_2 utilization processes are water stressed, which could lead to local competition for water resources.

> **RECOMMENDATION 4.6.** The U.S. Department of Energy (DOE) should work with its national laboratories to analyze the effect of CO_2 utilization, and the required enabling inputs (e.g., hydrogen), on local water demands. This analysis should identify regions that may experience water stress and opportunities for water infrastructure to serve multiple projects or facilities, while at the same time considering local environmental and justice impacts. The analysis should also explore the feasibility of using non-freshwater for some appli-

cations and deploying desalination projects that can provide water for hydrogen and other uses. The results of this analysis should be taken into account in project and location selection for any DOE-funded activity.

FINDING 4.15 Climate Impact of Decarbonizing Energy. In the long term, CO_2 utilization will fit into a larger system powered by net-zero-carbon-emissions energy, including decarbonized electric power for transportation, buildings, and other industrial activities. In the short term, with limited availability of emissions-free electricity, decarbonizing electric power systems will have more impact on total emissions reduction than using that same electricity to directly enable CO_2 utilization for net-zero chemicals and materials production. Photosynthetic processes for biological CO_2 conversion can use energy from the sun directly, without requiring any clean electricity infrastructure for photosynthesis.

4.7 REFERENCES

Abbas, Z., T. Mezher, and M.R.M. Abu-Zahra. 2013. "CO_2 Purification. Part I: Purification Requirement Review and the Selection of Impurities Deep Removal Technologies." *International Journal of Greenhouse Gas Control* 16(August):324–334. https://doi.org/10.1016/j.ijggc.2013.01.053.

Abramson, E., D. McFarlane, and J. Brown. 2020. "Transport Infrastructure for Carbon Capture and Storage: Whitepaper on Regional Infrastructure for Midcentury Decarbonization." Minneapolis, MN: Great Plains Institute. https://www.betterenergy.org/wp-content/uploads/2020/06/GPI_RegionalCO2Whitepaper.pdf.

Abramson, E., E. Thomley, and D. McFarlane. 2022. *An Atlas of Carbon and Hydrogen Hubs for United States Decarbonization*. Minneapolis, MN: Great Plains Institute. https://scripts.betterenergy.org/CarbonCaptureReady/GPI_Carbon_and_Hydrogen_Hubs_Atlas.pdf.

ACER (European Union Agency for the Cooperation of Energy Regulators). 2021. *Transporting Pure Hydrogen by Repurposing Existing Gas Infrastructure: Overview of Existing Studies and Reflections on the Conditions for Repurposing*. Ljubljana, Slovenia: ACER. https://acer.europa.eu/Official_documents/Acts_of_the_Agency/Publication/Transporting%20Pure%20Hydrogen%20by%20Repurposing%20Existing%20Gas%20Infrastructure_Overview%20of%20studies.pdf.

Al Baroudi, H., A. Awoyomi, K. Patchigolla, K. Jonnalagadda, and E.J. Anthony. 2021. "A Review of Large-Scale CO_2 Shipping and Marine Emissions Management for Carbon Capture, Utilisation and Storage." *Applied Energy* 287(April):116510. https://doi.org/10.1016/j.apenergy.2021.116510.

Alcalde, J., N. Heinemann, L. Mabon, R.H. Worden, H. de Coninck, H. Robertson, M. Maver, et al. 2019. "Acorn: Developing Full-Chain Industrial Carbon Capture and Storage in a Resource- and Infrastructure-Rich Hydrocarbon Province." *Journal of Cleaner Production* 233(October):963–971. https://doi.org/10.1016/j.jclepro.2019.06.087.

Allred, B.W., W.K. Smith, D. Twidwell, J.H. Haggerty, S.W. Running, D.E. Naugle, and S.D. Fuhlendorf. 2015. "Ecosystem Services Lost to Oil and Gas in North America." *Science* 348(6233):401–402. https://doi.org/10.1126/science.aaa4785.

Anchondo, C., and E. Klump. 2020. "Petra Nova Is Closed: What It Means for Carbon Capture." *E&E News*, September 22. https://www.eenews.net/articles/petra-nova-is-closed-what-it-means-for-carbon-capture.

ANL (Argonne National Laboratory). 2020. "Available Water Remaining for the United States (AWARE-US) Model." *Energy Systems*, February 18. https://greet.es.anl.gov/aware#page-header.

Bazzanella, A.M., and F. Ausfelder. 2017. *Low Carbon Energy and Feedstock for the European Chemical Industry. Technology Study*. Germany: DECHEMA. https://dechema.de/dechema_media/Downloads/Positionspapiere/Technology_study_Low_carbon_energy_and_feedstock_for_the_European_chemical_industry-p-20002750.pdf.

Benquet, C., A. Knarvik, E. Gjernes, O.A. Hvidsten, E.R. Kleppe, and S. Akhter. 2021. "First Process Results and Operational Experience with CESAR1 Solvent at TCM with High Capture Rates (ALIGN-CCUS Project)." *Proceedings of the 15th Greenhouse Gas Control Technologies Conference*. Abu Dhabi, UAE. https://dx.doi.org/10.2139/ssrn.3814712.

Bjerketvedt, V.S., A. Tomasgard, and S. Roussanaly. 2022. "Deploying a Shipping Infrastructure to Enable Carbon Capture and Storage from Norwegian Industries." *Journal of Cleaner Production* 333(January):129586. https://doi.org/10.1016/j.jclepro.2021.129586.

Blanton, E.M., M.C. Lott, and K.N. Smith. 2021. *Investing in the US Natural Gas Pipeline System to Support Net-Zero Targets*. New York: Columbia University Center on Global Energy Policy. https://www.energypolicy.columbia.edu/sites/default/files/file-uploads/GasPipelines_CGEP_Report_081721.pdf.

Bostick, D. 2019. "Flue Gas Aerosol Pretreatment Technologies to Minimize Post-Combustion CO_2 Capture (PCC) Solvent Losses." Presented at the 2019 NETL CO_2 Capture Technology Meeting, Pittsburgh, PA. August 26. https://netl.doe.gov/sites/default/files/netl-file/D-Bostick-Linde-Aerosol-Pretreatment.pdf.

Bostick, D. 2022. "Flue Gas Aerosol Pretreatment Technologies to Minimize PCC Solvent Losses." In *2022 Compendium of Carbon Capture Technologies*. DOE Project No. FE0031592. Linde Gas North America, LLC. https://edx.netl.doe.gov/dataset/flue-gas-aerosol-pretreatment-technologies-to-minimize-pcc-solvent-losses.

Brownsort, P.A. 2019. *Briefing on Carbon Dioxide Specifications for Transport*. 1st Report of the Thematic Working Group on CO_2 Transport, Storage and Networks. European Union: CCUS Projects Network. https://www.ccusnetwork.eu/sites/default/files/TG3_Briefing-CO2-Specifications-for-Transport.pdf.

C2ES (Center for Climate and Energy Solutions). 2021. "U.S. State Greenhouse Gas Emissions Targets." https://www.c2es.org/document/greenhouse-gas-emissions-targets.

CATF (Clean Air Task Force). 2021. "US Carbon Capture Activity and Project Map." https://www.catf.us/ccsmapus.

CATF. 2022. "Potential Carbon Capture Projects Under 45Q Federal Tax Credit." https://docs.google.com/spreadsheets/d/115hsADg3ymy3lKBy4PBQRXz_MBknptqlRtlfuv79XV8/edit#gid=1540463113.

Cemnet. 2022. "Cement Plants Located in United States." https://www.cemnet.com/global-cement-report/country/united-states.

Chemical Engineering. 2010. "Supercritical CO_2: A Green Solvent. Chemical Engineering: Essentials for the CPI Professional." https://www.chemengonline.com/supercritical-co2-a-green-solvent/?printmode=1.

Choi, Y.S., S. Nesic, and D. Young. 2010. "Effect of Impurities on the Corrosion Behavior of CO_2 Transmission Pipeline Steel in Supercritical CO_2–Water Environments." *Environmental Science & Technology* 44(23):9233–9238. https://doi.org/10.1021/es102578c.

CO_2Quest. 2017. "Overviews: Experiments Video." CO_2Quest: Impact of the Quality of CO_2 on Storage and Transport. 2017. https://www.co2quest.eu/overviews/#tab-1-3-experiments-video.

Cole, W.J., D. Greer, P. Denholm, A.W. Frazier, S. Machen, T. Mai, N. Vincent, and S.F. Baldwin. 2021. "Quantifying the Challenge of Reaching a 100% Renewable Energy Power System for the United States." *Joule* 5(7):1732–1748. https://doi.org/10.1016/j.joule.2021.05.011.

Concawe. 2021. "Technology Scouting—Carbon Capture: From Today's to Novel Technologies." *Concawe Review* 29(2):4–14.

Corsten, M., A. Ramírez, L. Shen, J. Koornneef, and A. Faaij. 2013. "Environmental Impact Assessment of CCS Chains—Lessons Learned and Limitations from LCA Literature." *International Journal of Greenhouse Gas Control* 13(March):59–71. https://doi.org/10.1016/j.ijggc.2012.12.003.

Deng, H., S. Roussanaly, and G. Skaugen. 2019. "Techno-Economic Analyses of CO_2 Liquefaction: Impact of Product Pressure and Impurities." *International Journal of Refrigeration* 103(July):301–315. https://doi.org/10.1016/j.ijrefrig.2019.04.011.

Devitt, D.A., M.H. Young, and J.P. Pierre. 2020. "Assessing the Potential for Greater Solar Development in West Texas, USA." *Energy Strategy Reviews* 29(May):100490. https://doi.org/10.1016/j.esr.2020.100490.

Digdaya, I.A., I. Sullivan, M. Lin, L. Han, W. Cheng, H.A. Atwater, and C. Xiang. 2020. "A Direct Coupled Electrochemical System for Capture and Conversion of CO_2 from Oceanwater." *Nature Communications* 11(1):4412. https://doi.org/10.1038/s41467-020-18232-y.

DOE (U.S. Department of Energy). 2022. "Biden-Harris Administration Launches $2.6 Billion Funding Programs to Slash Carbon Emissions." Energy.gov. https://www.energy.gov/articles/biden-harris-administration-launches-26-billion-funding-programs-slash-carbon-emissions?utm_medium=email&utm_source=govdelivery.

DOE-BETO (Bioenergy Technologies Office). 2021. *2021 Project Peer Review*. Washington, DC: DOE. https://www.energy.gov/sites/default/files/2022-06/beto-00-2021-peer-review-report.pdf.

DOE-EERE (Office of Energy Efficiency & Renewable Energy). 2017. "Confronting the Duck Curve: How to Address Over-Generation of Solar Energy." *Energy.gov*, October 12. https://www.energy.gov/eere/articles/confronting-duck-curve-how-address-over-generation-solar-energy.

EC (European Commission, Directorate-General for Climate Action). 2011. *Implementation of Directive 2009/31/EC on the Geological Storage of Carbon Dioxide: Guidance Document 2, Characterisation of the Storage Complex, CO_2 Stream Composition, Monitoring and Corrective Measures*. Luxembourg: Publications Office. https://doi.org/DOI:%20 10.2834/98293.

EEA (European Environment Agency). 2020. "Carbon Capture and Storage Could Also Impact Air Pollution." https://www.eea.europa.eu/highlights/carbon-capture-and-storage-could.

EFI (Energy Futures Initiative). 2021. *The Future of Clean Hydrogen in the United States: Views from Industry, Market Innovators, and Investors. From Kilograms to Gigatons: Pathways for Hydrogen Market Formation in the United States*. Washington, DC: EFI. https://energyfuturesinitiative.org/wp-content/uploads/sites/2/2022/03/The-Future-of-Clean-Hydrogen-in-the-U.S._Report-1.pdf.

EIA (U.S. Energy Information Administration). 2022a. "Energy-Related CO_2 Emission Data Tables." https://www.eia.gov/environment/emissions/state.

EIA. 2022b. "Natural Gas Explained: Liquefied Natural Gas." https://www.eia.gov/energyexplained/natural-gas/liquefied-natural-gas.php.

EIA. 2022c. "U.S. Liquefaction Capacity." https://www.eia.gov/naturalgas/data.php#imports.

EIGA (European Industrial Gases Association). 2016. "Carbon Dioxide Food and Beverages Grade, Source Qualification, Quality Standards and Verification." EIGA Doc 70/17, revision of Doc 70/08. https://www.eiga.eu/ct_documents/doc070-pdf.

EIGA. 2018. "Minimum Specifications for Food Gas Applications." EIGA Doc 126/20, revision of Doc 126/18. https://www.eiga.eu/ct_documents/doc126-pdf.

Eisaman, M.D., J.L.B. Rivest, S.D. Karnitz, C. de Lannoy, A. Jose, R.W. DeVaul, and K. Hannun. 2018. "Indirect Ocean Capture of Atmospheric CO_2: Part II. Understanding the Cost of Negative Emissions." *International Journal of Greenhouse Gas Control* 70(March):254–261. https://doi.org/10.1016/j.ijggc.2018.02.020.

Element Energy. 2018. *Shipping CO_2–UK Cost Estimation Study. Final report for Business, Energy & Industrial Strategy Department*. Cambridge, UK. https://assets.publishing.service.gov.uk/government/uploads/system/uploads/attachment_data/file/761762/BEIS_Shipping_CO2.pdf.

Eller, D. 2022. "What We Know About Three Carbon Capture Pipelines Proposed in Iowa." *Des Moines Register* April 24. https://www.desmoinesregister.com/story/money/business/2021/11/28/what-is-carbon-capture-pipeline-proposals-iowa-ag-ethanol-emissions/8717904002.

EPA (U.S. Environmental Protection Agency). 2000. "Appendix B–Overview of Acute Health Effects." In *Carbon Dioxide as a Fire Suppressant: Examining the Risks*. Washington, DC: Office of Air and Radiation Stratospheric Protection Division. https://www.epa.gov/sites/default/files/2015-06/documents/co2appendixb.pdf.

EPA. 2022a. "Summary Data: EGRID2020." https://www.epa.gov/system/files/documents/2022-01/egrid2020_summary_tables.pdf.

EPA. 2022b. "Understanding Global Warming Potentials." https://www.epa.gov/ghgemissions/understanding-global-warming-potentials.

Equinor. 2019. *Northern Lights Project Concept Report*. RE-PM673-00001. Washington, DC. https://ccsnorway.com/app/uploads/sites/6/2020/05/Northern-Lights-Project-Concept-report.pdf.

Ferrari, N., L. Mancuso, K. Burnard, and F. Consonni. 2019. "Effects of Plant Location on Cost of CO_2 Capture." *International Journal of Greenhouse Gas Control* 90(November):102783. https://doi.org/10.1016/j.ijggc.2019.102783.

Field, R.A., and R.G. Derwent. 2021. "Global Warming Consequences of Replacing Natural Gas with Hydrogen in the Domestic Energy Sectors of Future Low-Carbon Economies in the United Kingdom and the United States of America." *International Journal of Hydrogen Energy* 46(58):30190–30203. https://doi.org/10.1016/j.ijhydene.2021.06.120.

Geske, J., N. Berghout, and M. van den Broek. 2015. "Cost-Effective Balance between CO_2 Vessel and Pipeline Transport. Part I—Impact of Optimally Sized Vessels and Fleets." *International Journal of Greenhouse Gas Control* 36(May):175–188. https://doi.org/10.1016/j.ijggc.2015.01.026.

Gibbins, J., and M. Lucquiaud. 2021. *BAT Review for New-Build and Retrofit Post-Combustion Carbon Dioxide Capture Using Amine-Based Technologies for Power and CHP Plants Fueled by Gas and Biomass as an Emerging Technology Under the IED for the UK, UKCCSRC Report*. UKCCSRC Report, ver.1.0. Sheffield, UK: UK CCS Research Centre. https://ukccsrc.ac.uk/wp-content/uploads/2021/06/BAT-for-PCC_V1_0.pdf.

Global CCS Institute. 2022. "CO_2RE Facilities Database." https://co2re.co/FacilityData.

Gorset, O., J.N. Knudsen, O.M. Bade, and I. Askestad. 2014. "Results from Testing of Aker Solutions Advanced Amine Solvents at CO_2 Technology Centre Mongstad." *Energy Procedia* 63:6267–6280. https://doi.org/10.1016/j.egypro.2014.11.658.

Green Hydrogen Coalition. n.d. "HyDeal Los Angeles: Architecting the Green Hydrogen Ecosystem for a Deeply Decarbonized LA." https://static1.squarespace.com/static/5e8961cdcbb9c05d73b3f9c4/t/6179eb9cf8ac24238842374d/1635380127410/HyDeal+LA+Phase+1+Takeaways.pdf.

Gulf Publishing Holdings LLC. 2022. "US to Drive LNG Liquefaction Capacity Growth in North America by 2024." *Gas Processing & LNG*. http://www.gasprocessingnews.com/news/us-to-drive-lng-liquefaction-capacity-growth-in-north-america-by-2024.aspx.

Hewitt, D. 2019. "Behold the Butterfly Graph. Northeast Energy Efficiency Partnerships." https://neep.org/blog/behold-butterfly-graph.

Hiestermann, M. 2022. "Hydrogen Supply Chain Portfolios in the Netherlands Between 2030 and 2050." Master thesis, Delft University of Technology, Netherlands. https://repository.tudelft.nl/islandora/object/uuid%3Ac4b1c997-f0f6-4859-ba9b-0f579c8d9052.

HM Government. 2021. *Industrial Decarbonisation Strategy*. United Kingdom: Department for Business, Energy & Industrial Strategy. https://assets.publishing.service.gov.uk/government/uploads/system/uploads/attachment_data/file/970229/Industrial_Decarbonisation_Strategy_March_2021.pdf.

Ho, E. 2006. *Elastomeric Seals for Rapid Gas Decompression Applications in High-Pressure Services*. Bedfordshire, UK: BHR Group Limited. https://www.hse.gov.uk/research/rrpdf/rr485.pdf.

Holmgren, J. 2022. "Infrastructure Needs for Capture and Utilization Companies: LanzaTech." Presentation to the National Academies' Committee on Carbon Utilization Infrastructure, Markets, Research and Development Meeting #2. March 2. https://www.nationalacademies.org/event/03-01-2022/carbon-utilization-infrastructure-markets-research-and-development-meeting-2.

Hopcroft, T., and A. Papadamou. 2021. "The Water Industry in a Hydrogen Economy." *PA Consulting*, July 12. https://www.paconsulting.com/newsroom/expert-opinion/the-water-report-the-water-industry-in-a-hydrogen-economy-12-july-2021.

IEA (International Energy Agency). 2019. *The Future of Hydrogen: Seizing Today's Opportunities*. Paris. https://iea.blob.core.windows.net/assets/9e3a3493-b9a6-4b7d-b499-7ca48e357561/The_Future_of_Hydrogen.pdf.

IEA. 2020. *Energy Technology Perspectives 2020 – Special Report on Carbon Capture Utilisation and Storage: CCUS in Clean Energy Transitions*. Paris. https://iea.blob.core.windows.net/assets/181b48b4-323f-454d-96fb-0bb1889d96a9/CCUS_in_clean_energy_transitions.pdf.

IEA. 2021. "Is Carbon Capture Too Expensive?" https://www.iea.org/commentaries/is-carbon-capture-too-expensive.

IEAGHG (IEA Greenhouse Gas R&D Programme). 2004. *Ship Transport of CO_2*. Report No. PH4/30. Cheltenham, UK. https://ieaghg.org/docs/General_Docs/Reports/PH4-30%20Ship%20Transport.pdf.

IEAGHG. 2018. *Cost of CO_2 Capture in the Industrial Sector: Cement and Iron and Steel Industries. IEAGHG Technical Review 2018-TR03*. Cheltenham, UK. https://ieaghg.org/publications/technical-reports/reports-list/10-technical-reviews/931-2018-tr03-cost-of-co2-capture-in-the-industrial-sector-cement-and-iron-and-steel-industries.

IRENA (International Renewable Energy Agency). 2022. *Geopolitics of the Energy Transformation: The Hydrogen Factor*. Abu Dhabi: IRENA. https://www.irena.org/-/media/Files/IRENA/Agency/Publication/2022/Jan/IRENA_Geopolitics_Hydrogen_2022.pdf.

Jackson, S., and E. Brodal. 2019. "Optimization of the CO_2 Liquefaction Process-Performance Study with Varying Ambient Temperature." *Applied Sciences* 9(20):4467. https://doi.org/10.3390/app9204467.

Jenkins, J. 2022. "CO_2 Transportation Options." Presentation to the National Academies' Committee on Carbon Utilization Infrastructure, Markets, Research and Development Meeting #2. https://www.nationalacademies.org/event/03-01-2022/carbon-utilization-infrastructure-markets-research-and-development-meeting-2.

Kearns, D., H. Liu, and C. Consoli. 2021. *Technology Readiness and Costs of CCS*. The Circular Carbon Economy: Keystone to Global Sustainability Series. Melbourne, AU: Global CCS Institute. https://www.globalccsinstitute.com/wp-content/uploads/2021/03/Technology-Readiness-and-Costs-for-CCS-2021-1.pdf.

Keith, D.W., G. Holmes, D. St. Angelo, and K. Heidel. 2018. "A Process for Capturing CO_2 from the Atmosphere." *Joule* 2(8):1573–1594. https://doi.org/10.1016/j.joule.2018.05.006.

Khan, M.A., T. Al-Attas, S. Roy, M.M. Rahman, N. Ghaffour, V. Thangadurai, S. Larter, J. Hu, P.M. Ajayan, and M.G. Kibria. 2021. "Seawater Electrolysis for Hydrogen Production: A Solution Looking for a Problem?" *Energy & Environmental Science* 14(9):4831–4839. https://doi.org/10.1039/D1EE00870F.

Knoope, M.M.J., I.M.E. Raben, A. Ramírez, M.P.N. Spruijt, and A.P.C. Faaij. 2014. "The Influence of Risk Mitigation Measures on the Risks, Costs and Routing of CO_2 Pipelines." *International Journal of Greenhouse Gas Control* 29(October):104–124. https://doi.org/10.1016/j.ijggc.2014.08.001.

Lampert, D.J., H. Cai, Z. Wang, J. Keisman, M. Wu, J. Han, J. Dunn, et al. 2015. *Development of a Life Cycle Inventory of Water Consumption Associated with the Production of Transportation Fuels*. ANL/ESD-15/27. Argonne, IL: Argonne National Laboratory. https://doi.org/10.2172/1224980.

Lampert, D.J., H. Cai, and A. Elgowainy. 2016. "Wells to Wheels: Water Consumption for Transportation Fuels in the United States." *Energy & Environmental Science* 9(3):787–802. https://doi.org/10.1039/C5EE03254G.

Larson, E., C. Greig, J. Jenkins, E. Mayfield, A. Pascale, C. Zhang, J. Drossman, et al. 2021. *Net-Zero America: Potential Pathways, Infrastructure, and Impacts*. Final Report. Princeton, NJ: Princeton University. https://netzeroamerica.princeton.edu/the-report.

LBNL (Lawrence Berkeley National Laboratory). 2022. "Generation, Storage, and Hybrid Capacity in Interconnection Queues." https://emp.lbl.gov/generation-storage-and-hybrid-capacity.

Lebling, K., H. Leslie-Bole, Z. Byrum, and E. Bridgwater. 2022. "6 Things to Know About Direct Air Capture. World Resources Institute." https://www.wri.org/insights/direct-air-capture-resource-considerations-and-costs-carbon-remova.

Lee, U., Y. Lim, S. Lee, J. Jung, and C. Han. 2012. "CO_2 Storage Terminal for Ship Transportation." *Industrial & Engineering Chemistry Research* 51(1):389–397. https://doi.org/10.1021/ie200762f.

Linde. 2022. "Oxygen-Enhanced Combustion for Metal Production." https://www.lindeus.com/industries/metal-production/oxygen-enhanced-combustion.

Liu, N. 2021. "Increasing Blue Hydrogen Production Affordability." *Hydrocarbon Processing* June. https://www.hydrocarbonprocessing.com/magazine/2021/june-2021/special-focus-process-optimization/increasing-blue-hydrogen-production-affordability.

Mac Dowell, N., N. Sunny, N. Brandon, H. Herzog, A.Y. Ku, W. Maas, A. Ramirez, D.M. Reiner, G.N. Sant, and N. Shah. 2021. "The Hydrogen Economy: A Pragmatic Path Forward." *Joule* 5(10):2524–2529, reprint. https://doi.org/10.1016/j.joule.2021.09.014.

Mahgerefteh, H., S. Brown, and G. Denton. 2012. "Modelling the Impact of Stream Impurities on Ductile Fractures in CO_2 Pipelines." *Chemical Engineering Science* 74(May):200–210. https://doi.org/10.1016/j.ces.2012.02.037.

Martynov, S.B., R.H. Talemi, S. Brown, and H. Mahgerefteh. 2017. "Assessment of Fracture Propagation in Pipelines Transporting Impure CO_2 Streams." *Energy Procedia* 114 (July):6685–6697. https://doi.org/10.1016/j.egypro.2017.03.1797.

McKaskle, R., K. Fisher, P. Selz, and Y. Lu. 2018. *Evaluation of Carbon Dioxide Capture Options from Ethanol Plants.* Circular 595. Champaign: Illinois State Geological Survey and Prairie Research Institute. https://library.isgs.illinois.edu/Pubs/pdfs/circulars/c595.pdf.

Myers, J.R., G.S. Ruberto, R.A. Paccella, J.A. Ventura, A.L. Boehman, R.J. Briggs, M.T. Pietrucha, S. Bloser, and J.R. Anstrom. 2012. *Feasibility Study for Liquefied Natural Gas Utilization for Commercial Vehicles on the Pennsylvania Turnpike.* PSU-2011-03. Mid-Atlantic Universities Transportation Center. https://rosap.ntl.bts.gov/view/dot/25519.

Naraharisetti, P.K., T.Y. Yeo, and J. Bu. 2017. "Factors Influencing CO_2 and Energy Penalties of CO_2 Mineralization Processes." *ChemPhysChem* 18(22):3189–3202. https://doi.org/10.1002/cphc.201700565.

NASEM (National Academies of Sciences, Engineering, and Medicine). 2019. *Gaseous Carbon Waste Streams Utilization: Status and Research Needs.* Washington, DC: The National Academies Press. https://doi.org/10.17226/25232.

NASEM. 2021. *The Future of Electric Power in the United States.* Washington, DC: The National Academies Press. https://doi.org/10.17226/25968.

Neele, F., R. de Kler, M. Nienoord, P. Brownsort, J. Koornneef, S. Belfroid, L. Peters, A. van Wijhe, and D. Loeve. 2017. "CO_2 Transport by Ship: The Way Forward in Europe." *Energy Procedia* 114:6824–6834. https://doi.org/10.1016/j.egypro.2017.03.1813.

NETL (National Energy Technology Laboratory). 2015. *A Review of the CO_2 Pipeline Infrastructure in the U.S. DOE/NETL-2014/1681.* Washington, DC: U.S. Department of Energy. https://www.energy.gov/sites/prod/files/2015/04/f22/QER%20Analysis%20-%20A%20Review%20of%20the%20CO2%20Pipeline%20Infrastructure%20in%20the%20U.S_0.pdf.

NETL. 2019. *CO_2 Impurity Design Parameters. Quality Guidelines for Energy System Studies.* Pittsburgh: National Energy Technology Laboratory. https://www.netl.doe.gov/projects/files/QGESSCO2ImpurityDesignParameters_010119.pdf.

NETL. 2020. *2020 Carbon Capture Program R&D: Compendium of Carbon Capture Technology.* Pittsburgh: National Energy Technology Laboratory. https://www.netl.doe.gov/sites/default/files/2020-07/Carbon-Capture-Technology-Compendium-2020.pdf.

NETL. 2022. "Carbon Storage Program." U.S. Department of Energy. https://netl.doe.gov/sites/default/files/2022-08/Program-116.pdf.

New Jersey Department of Health. 2016. "Hazardous Substance Fact Sheet: Carbon Dioxide." https://nj.gov/health/eoh/rtkweb/documents/fs/0343.pdf.

Nickel, R., L. Hampton, and N. Williams. 2022. "N. America's Old Pipelines Seek New Life Moving Carbon in Climate Push." *Reuters* February 23. https://www.reuters.com/business/sustainable-business/n-americas-old-pipelines-seek-new-life-moving-carbon-climate-push-2022-02-23.

NPC (National Petroleum Council). 2021. "CO_2 Transport." Chapter 6 in *Meeting the Dual Challenge: A Roadmap to At-Scale Deployment of Carbon Capture, Use, and Storage.* Washington, DC. https://dualchallenge.npc.org/files/CCUS-Chap_6-030521.pdf.

Ozkan, M. 2021. "Direct Air Capture of CO_2: A Response to Meet the Global Climate Targets." *MRS Energy & Sustainability* 8(2):51–56. https://doi.org/10.1557/s43581-021-00005-9.

Ozkan, M., S.P. Nayak, A.D. Ruiz, and W. Jiang. 2022. "Current Status and Pillars of Direct Air Capture Technologies." *IScience* 25(4):103990. https://doi.org/10.1016/j.isci.2022.103990.

Peletiri, S.P., N. Rahmanian, and I.M. Mujtaba. 2018. "CO_2 Pipeline Design: A Review." *Energies* 11(9):2184. https://doi.org/10.3390/en11092184.

Peletiri, S.P., I.M. Mujtaba, and N. Rahmanian. 2019. "Process Simulation of Impurity Impacts on CO_2 Fluids Flowing in Pipelines." *Journal of Cleaner Production* 240(December):118145. https://doi.org/10.1016/j.jclepro.2019.118145.

Penev, M., N. Rustagi, C. Hunter, and J. Eichman. 2019. "Energy Storage: Days of Service Sensitivity Analysis." Presentation to the Hydrogen and Fuel Cell Technical Advisory Committee. March 19. https://www.nrel.gov/docs/fy19osti/73520.pdf.

REPEAT (Rapid Energy Policy Evaluation and Analysis Toolkit). 2022. "Repeatproject." repeatproject.org.

Reyes-Lúa, A., Y. Arellano, I. Treu Røe, L. Rycroft, T. Wildenborg, and K. Jordal. 2021. *CO_2 Ship Transport: Benefits for Early Movers and Aspects to Consider*. ENER/C2/2017-65/SI2.793333. European Union: CCUS Projects Network. https://www.ccusnetwork.eu/sites/default/files/TG3_Briefing-CO2-ship-transport-Benefits-for-early-movers-and-aspects-to-consider.pdf.

Rosa, L., D.L. Sanchez, G. Realmonte, D. Baldocchi, and P. D'Odorico. 2021. "The Water Footprint of Carbon Capture and Storage Technologies." *Renewable and Sustainable Energy Reviews* 138(March):110511. https://doi.org/10.1016/j.rser.2020.110511.

Roussanaly, S., H. Deng, G. Skaugen, and T. Gundersen. 2021. "At What Pressure Shall CO_2 Be Transported by Ship? An in-Depth Cost Comparison of 7 and 15 Barg Shipping." *Energies* 14(18):5635. https://doi.org/10.3390/en14185635.

Rustagi, N. 2022. "NREL Systems Analysis Overview." Presentation at the DOE Hydrogen Annual Merit Review. June 6. https://www.hydrogen.energy.gov/pdfs/review22/plenary9_rustagi_2022_o.pdf.

Ruth, M.F., P. Jadun, N. Gilroy, E. Connelly, R. Boardman, A.J. Simon, A. Elgowainy, and J. Zuboy. 2020. *The Technical and Economic Potential of the H_2@Scale Concept Within the United States*. NREL/TP-6A20-77610. Golden, CO: National Renewable Energy Laboratory. https://www.nrel.gov/docs/fy21osti/77610.pdf.

Rütters, H., D. Bettge, R. Eggers, A. Kather, C. Lempp, U. Lubenau, and COORAL-Team. 2015. *CO_2-Reinheit Für Die Abscheidung Und Lagerung (COORAL)–Synthese*. Hannover: Bundesanstalt für Geowissenschaften und Rohstoffe. https://www.bgr.bund.de/DE/Themen/Nutzung_tieferer_Untergrund_CO2Speicherung/CO2Speicherung/COORAL/Downloads/Synthesebericht.pdf?__blob=publicationFile&v=6.

Rystad Energy. 2021. "Green Hydrogen Projects Will Stay Dry Without a Parallel Desalination Market to Provide Fresh Water." https://www.hellenicshippingnews.com/green-hydrogen-projects-will-stay-dry-without-a-parallel-desalination-market-to-provide-fresh-water.

Shaw, D. 2009. "Cansolv CO_2 Capture: The Value of Integration." *Energy Procedia* 1(1):237–246. https://doi.org/10.1016/j.egypro.2009.01.034.

Shi, X., X. Liao, and Y. Li. 2020. "Quantification of Fresh Water Consumption and Scarcity Footprints of Hydrogen from Water Electrolysis: A Methodology Framework." *Renewable Energy* 154(July):786–796. https://doi.org/10.1016/j.renene.2020.03.026.

SINTEF. 2019. "CO_2 Impurities: What Else Is There in CO_2 except CO_2?" Blog. November 6. https://blog.sintef.com/sintefenergy/energy-efficiency/what-else-is-there-in-co2-except-co2.

Smith, W.A., T. Burdyny, D.A. Vermaas, and H. Geerlings. 2019. "Pathways to Industrial-Scale Fuel Out of Thin Air from CO_2 Electrolysis." *Joule* 3(8):1822–1834. https://doi.org/10.1016/j.joule.2019.07.009.

Soraghan, M., and C. Anchondo. 2022. "Biden Releases Plan to Avoid "Dangerous" CO_2 Pipeline Failures." *E&E News*, May 27. https://www.eenews.net/articles/biden-releases-plan-to-avoid-dangerous-co2-pipeline-failures.

Spietz, T., S. Dobras, L. Więcław-Solny, and A. Krótki. 2017. "Nitrosamines and Nitramines in Carbon Capture Plants." *Environmental Protection and Natural Resources* 28(4):43–50. https://doi.org/10.1515/oszn-2017-0027.

Stolaroff, J.K., S.H. Pang, W. Li, W.G. Kirkendall, H.M. Goldstein, R.D. Aines, and S.E. Baker. 2021. "Transport Cost for Carbon Removal Projects with Biomass and CO_2 Storage." *Frontiers in Energy Research* 9(May). https://www.frontiersin.org/article/10.3389/fenrg.2021.639943.

Talemi, R.H., S. Brown, S. Martynov, and H. Mahgerefteh. 2016. "Hybrid Fluid–Structure Interaction Modelling of Dynamic Brittle Fracture in Steel Pipelines Transporting CO_2 Streams." *International Journal of Greenhouse Gas Control* 54(November):702–715. https://doi.org/10.1016/j.ijggc.2016.08.021.

Talemi, R., S. Cooreman, H. Mahgerefteh, S. Martynov, and S. Brown. 2019. "A Fully Coupled Fluid-Structure Interaction Simulation of Three-Dimensional Dynamic Ductile Fracture in a Steel Pipeline." *Theoretical and Applied Fracture Mechanics* 101(June):224–235. https://doi.org/10.1016/j.tafmec.2019.02.005.

Tarroja, B., F. Chiang, A. AghaKouchak, and S. Samuelsen. 2018. "Assessing Future Water Resource Constraints on Thermally Based Renewable Energy Resources in California." *Applied Energy* 226(September):49–60. https://doi.org/10.1016/j.apenergy.2018.05.105.

Tollefson, J. 2018. "Sucking Carbon Dioxide from Air Is Cheaper Than Scientists Thought." *Nature* 558(173). https://doi.org/10.1038/d41586-018-05357-w.

UNFCCC (United Nations Framework Convention on Climate Change). 2022. "The Paris Agreement." https://unfccc.int/process-and-meetings/the-paris-agreement/the-paris-agreement.

Van Atten, C., A. Saha, L. Hellgren, and T. Langlois. 2021. *Benchmarking Air Emissions of the 100 Largest Electric Power Producers in the United States*. Concord, MA: M.J. Bradley & Associates. https://www.mjbradley.com/sites/default/files/Presentation_of_Results_2021.pdf.

Van der Giesen, C., C.J. Meinrenken, R. Kleijn, B. Sprecher, K.S. Lackner, and G.J. Kramer. 2017. "A Life Cycle Assessment Case Study of Coal-Fired Electricity Generation with Humidity Swing Direct Air Capture of CO_2 Versus MEA-Based Post-combustion Capture." *Environmental Science & Technology* 51(2):1024–1034. https://doi.org/10.1021/acs.est.6b05028.

van Zalk, J., and P. Behrens. 2018. "The Spatial Extent of Renewable and Non-Renewable Power Generation: A Review and Meta-Analysis of Power Densities and Their Application in the U.S." *Energy Policy* 123(December):83–91. https://doi.org/10.1016/j.enpol.2018.08.023.

Wang, H., J. Chen, and Q. Li. 2019. "A Review of Pipeline Transportation Technology of Carbon Dioxide." *IOP Conference Series: Earth and Environmental Science* 310(3):032033. https://doi.org/10.1088/1755-1315/310/3/032033.

Warwick, N., P. Griffiths, J. Keeble, A. Archibald, J. Pyle, and K. Shine. 2022. *Atmospheric Implications of Increased Hydrogen Use*. Cambridge, UK: University of Cambridge, National Centre for Atmospheric Sciences, University of Reading. https://assets.publishing.service.gov.uk/government/uploads/system/uploads/attachment_data/file/1067144/atmospheric-implications-of-increased-hydrogen-use.pdf.

Wilday, J., J.L. Saw, M. Wardman, and M. Bilio. 2014. "The CO_2PipeHaz Good Practice Guidelines for CO_2 Pipeline Safety." Institution of Chemical Engineers Symposium Series, January. https://www.icheme.org/media/8946/xxiv-paper-51.pdf.

Woolley, R.M., M. Fairweather, C.J. Wareing, S.A.E.G. Falle, H. Mahgerefteh, S. Martynov, S. Brown, et al. 2014. "CO_2Pipe-Haz: Quantitative Hazard Assessment for Next Generation CO_2 Pipelines." *Energy Procedia* 63:2510–2529. https://doi.org/10.1016/j.egypro.2014.11.274.

Xu, H., U. Lee, A.M. Coleman, M.S. Wigmosta, and M. Wang. 2019. "Assessment of Algal Biofuel Resource Potential in the United States with Consideration of Regional Water Stress." *Algal Research* 37(January):30–39. https://doi.org/10.1016/j.algal.2018.11.002.

Yoshimura, R., S. Wai, Y. Ota, K. Nishioka, and Y. Suzuki. 2022. "Effects of Artificial River Water on PEM Water Electrolysis Performance." *Catalysts* 12(9):934. https://doi.org/10.3390/catal12090934.

Zahid, U., J. An, C. Lee, U. Lee, and C. Han. 2015. "Design and Operation Strategy of CO_2 Terminal." *Industrial & Engineering Chemistry Research* 54(8):2353–2365. https://doi.org/10.1021/ie503696x.

Zeng, Y., and K. Li. 2020. "Influence of SO_2 on the Corrosion and Stress Corrosion Cracking Susceptibility of Supercritical CO_2 Transportation Pipelines." *Corrosion Science* 165(April):108404. https://doi.org/10.1016/j.corsci.2019.108404.

ZEP (Zero Emissions Platform). 2011. *The Costs of CO_2 Transport: Post-Demonstration CCS in the EU*. Brussels: European Technology Platform for Zero Emission Fossil Fuel Power Plants. https://www.globalccsinstitute.com/archive/hub/publications/119811/costs-co2-transport-post-demonstration-ccs-eu.pdf.

ZEP. 2020. *A Trans-European Transportation Infrastructure for CCUS: Opportunities & Challenges*. Brussels: Zero Emissions Platform. https://zeroemissionsplatform.eu/wp-content/uploads/A-Trans-European-CO2-Transportation-Infrastructure-for-CCUS-Opportunities-Challenges-1.pdf.

Zhang, X., F. Jin, X. Yuan, and H. Zhang. 2021. "Low-Carbon Multimodal Transportation Path Optimization Under Dual Uncertainty of Demand and Time." *Sustainability* 13(15):8180. https://doi.org/10.3390/su13158180.

5

Policy, Regulatory, and Societal Considerations for CO_2 Utilization Systems

This chapter introduces policy, regulatory, and societal considerations relative to the expansion of carbon dioxide (CO_2) capture and utilization. The chapter starts by providing broad policy considerations and continues with a presentation of the current regulatory framework for CO_2 utilization, storage, and transportation, highlighting challenges and proposing solutions. Policies meant to expand the CO_2 utilization economy could have societal impacts that negatively affect already disadvantaged communities. The chapter suggests that the environmental justice framework can help constructively reveal, manage, and resolve societal concerns.

5.1 POLICY AND REGULATORY CONSIDERATIONS

This section starts by laying out general guidelines for efficient regulation of CO_2 utilization. Cost-benefit analysis (CBA) rules can be applied to prioritize projects when regulators work with a predetermined budget. Emphasis is on policies that deal with both environmental and knowledge externalities, but potential economies of scale call for targeted policies. Regulation needs to be framed to reduce uncertainty, for example, from regulatory gaps, and must avoid unnecessary bureaucratic costs for private investors.

5.1.1 Cost-Effective Regulation of Environmental and Knowledge Externalities

In the current policy environment, insufficient investments in low-carbon technologies, emissions abatement, CO_2 utilization, and the supporting infrastructure are at least in part due to insufficient incentives for investors to make costly investments that have global benefits much larger than private costs (Nordhaus 2019). In other words, absent regulation, investors and consumers do not prioritize the societal cost of emitting CO_2 and other greenhouse gases (GHGs) when making their investment and consumption choices. Societal costs that are not borne by investors and consumers are called negative externalities (Tietenberg and Lewis 2018).

This misalignment between private and societal costs leads to a level of CO_2 emissions that generates more harm than good for society as a whole. For example, in the specific case of CO_2, the damage caused by emitting one additional ton of CO_2 is close to zero for the emitter, while the cost for the entire planet, also including future generations, is estimated to be equal to $50 per ton of CO_2 or higher (IWG 2021).

Similarly, individuals and firms invest in research and development (R&D), adoption, and diffusion[1] of technology less than what would be beneficial for society as a whole because they do not consider the full benefit of their private investment (Arrow 1962; Griliches 1957; Samuelson 1954). For example, if firms that invest in R&D cannot fully protect their innovation, a fraction of their investment has benefits for society but cannot be monetized. This pushes firms to invest less than what would be optimal for society as a whole (Romer 1990). The greater the extent of the public knowledge spillover, the greater the gap between private and socially optimal investment. This explains the important role of public grants for basic research that typically has large spillovers. Without government intervention, there would be much less innovation in science. Knowledge spillovers are an example of positive externalities. Similar problems arise when investment in new technologies generates cost reductions due to learning effects. Early investors pay a higher price, but late investors benefit from cost reductions. Thus, there is an incentive to wait for cost reductions, which leads to a slower adoption of the technology and slower cost reductions than if early investors were rewarded for their contribution to society. This is particularly important when early investments also reduce risk by generating more information on the new technology. For example, learning-by-doing explains a large fraction of the observed cost reduction in solar photovoltaic (PV) technology (Kavlak et al. 2018). Early investors in CO_2 utilization technologies will face similar burdens. Without investments that lead to societal benefits and a policy framework that rewards knowledge creation, investments in R&D and in new technologies are still possible but not as large as they otherwise could be.

The CO_2 utilization economy is currently at a disadvantage against incumbent products because present policies do not penalize CO_2 emissions and do not reward innovation and technology diffusion as it would be optimal to do. Ideally, policy would address simultaneously both negative and positive externalities.

For example, production of hydrocarbons from CO_2 will compete with production of those same units from fossil carbon (or biomass). At today's costs of oil, CO_2, and hydrogen, in today's regulatory environment, and with today's technology, fossil carbon is the predominant source of hydrocarbon chemicals. Figure 5-1 shows the

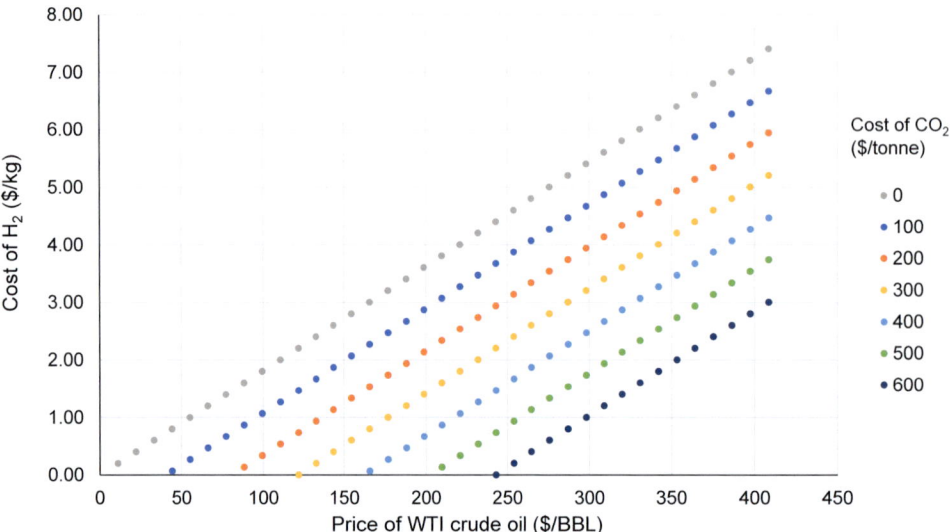

FIGURE 5-1 Maximum cost of hydrogen input that allows synthetic hydrocarbon fuels and other hydrocarbon chemicals to be cost competitive with oil-derived hydrocarbon fuels and chemicals, at different prices of crude oil (WTI = West Texas Intermediate crude oil) and for different CO_2 input costs. Assumes further reaction processing (capital expenditure, energy requirements) to make petrochemicals is similar.

---

[1] Technology diffusion is defined as the process by which the market for a new technology changes over time and from which production and usage patterns of new products and production processes result (Stoneman and Battisti, 2010).

approximate feedstock cost to transform CO_2 and hydrogen into synthetic hydrocarbon products versus deriving hydrocarbons from petroleum.

Even with a future hydrogen cost of $1/kg hydrogen and a reduction in the cost of CO_2 suitable for Track 2 short-lived products (such as direct air capture [DAC]-derived CO_2) to around $250/tonne CO_2, sustainable pathways for CO_2 utilization remain challenging. The limited point-source, non-fossil CO_2 available at low cost ($40/tonne) from bioethanol production is attractive at a low hydrogen price for synthetic fuels that competes with oil at $100/bbl; this is not true when oil is $50/bbl. This analysis excludes the option of making oxygenated products such as glycols from CO_2 and hydrogen, that is, incorporating oxygen from CO_2 into final products, which is an advantage of starting with CO_2 or biomass-based carbon feedstocks versus the fully reduced carbon in petroleum. CO_2 and hydrogen costs are not likely to diminish via further technology development to the point where these feedstocks will be more competitive than crude oil without an imposed use penalty on fossil carbon or an emissions penalty on fossil-derived CO_2 emissions.

It is possible to use a range of regulatory tools to penalize CO_2 emissions from traditional products (Tietenberg and Lewis 2018). At one end of the spectrum, "command-and-control" policies impose detailed and extensive rules that constrain private behavior to implement the desired solution for society. For example, the government can impose specific targets in terms of CO_2 utilization technology adoption. At the other end of the spectrum, regulation does not prescribe a specific behavior, or technology, but imposes sanctions on emissions. One way to do the latter is by imposing a carbon tax. Alternatively, an emission trading scheme, in which the price of carbon emerges from trading a fixed number of permits, can similarly achieve the same goal (Goulder and Schein 2013; Parry et al. 2022; Stavins 2019). In either policy case, transparency for consumers can be enhanced by the creation of a government-backed program to communicate the carbon intensity of a product. Similar to the Energy Star program (Energy Star 2022), a Carbon Star program could provide simple, credible, and unbiased information on the life cycle carbon content associated with a particular product.

Cost-effective (least cost) regulatory mixes ensure that marginal abatement costs are equal across all technologies and sectors (Tietenberg and Lewis 2018). Carbon pricing—using either carbon taxes or emissions trading—has the advantage over command-and-control in that it is cost-effective by design, because it puts all technologies on a level playing field while leaving investors and consumers the flexibility to choose the most convenient options (Tietenberg and Lewis 2018). Command-and-control regulation can be cost-effective, but it is usually not because regulators do not typically have all the information to choose the least-cost technology mix (Tietenberg and Lewis 2018).

While command-and-control policies can be more politically attractive, regulators need to carefully assess the added costs—or the missed mitigation potential—compared to carbon pricing. A large disparity of marginal abatement costs across technologies, sectors, and policies can be a signal of large cost inefficiencies. In the specific case of the United States, the advantages of a federal carbon price have been extensively documented in a large literature, and recently by Goulder and Hafstead (2018), Metcalf (2019), Parry et al. (2015), and Stavins (2020).

To enable optimal investment in technologies with significant knowledge spillovers, a second policy tool is needed. Subsidies—including grants for research, tax credits, and direct government purchases—can be used to stimulate R&D and adoption of CO_2 utilization technologies with positive knowledge externalities. To a large extent, the two externalities can be treated separately, and R&D support in carbon utilization technologies can be given equal weight to R&D support in other sectors of the economy (Nordhaus 2011).

Subsidies have to be calibrated to reflect the benefits of knowledge spillovers from R&D investment or from technology adoption, which is admittedly a difficult task. A growing empirical literature estimates the size of both knowledge spillovers and learning-by-doing effects in low-carbon energy technologies (Aghion et al. 2016; Popp 2002; Verdolini and Galeotti 2011). In general, societal returns to R&D are estimated to be two to three times the private returns (Mansfield 1996; Nordhaus 2011). Learning effects have been estimated for many energy technologies (Blanco et al. 2022), with central tendencies around 20 percent cost reduction for each doubling of deployment (McDonald and Schrattenholzer 2001). More information is needed to estimate the potential cost reductions in CO_2 utilization technologies.

Note that preferable use of subsidies in place of taxes or other policies aimed at reducing emissions is not optimal. Subsidies to finance reduction of emissions—such as tax credits for emissions reductions—are not efficient in the sense that they are costlier for society than, for example, carbon pricing. Subsidies for emissions reduction may be the only politically feasible policy tool, but regulators need to be aware of their cost.

A subsidy for emissions reductions makes consuming or producing the polluting goods artificially cheaper than what it would be with carbon pricing. Without additional constraints, this results in excessive production or consumption. For example, a tax on gasoline increases the cost of driving, pushing some consumers to walk and reducing overall number of miles traveled, reflecting societal costs and benefits, while a subsidy to clean fuels that keeps the cost of driving unchanged does not have the effect of reducing driving. This occurs because the cost of the subsidy is borne by society at large, while the tax on gasoline is borne by the individual user. Another problem with subsidies is that they must be financed with either higher taxes or lower government spending, both of which generate net welfare losses for the community (Parry 1998). Carbon pricing can instead be used to raise revenue that can finance tax cuts that can stimulate the economy and employment (Bovenberg 1999; Bovenberg and Goulder 1994; Carraro et al. 1996; Goulder 1995; Parry 1995). Carbon tax revenues also can be used to alleviate the likely regressive impact of energy taxes on disadvantaged groups with direct transfers (Budolfson et al. 2021; Goulder et al. 2018). Finally, subsidies are particularly cost-ineffective if they require the use of specific technologies—as for other command-and-control regulation—because the solution chosen by the regulator may not be optimal for many investors and consumers.

In the context of CO_2 utilization, these policy indications suggest that the most cost-effective policy framework relies on an economy-wide carbon pricing signal that is not limited to traditional competitors of CO_2 utilization products (i.e., fossil-fuel–derived chemicals and materials). Targeted interventions need to be directed to support fundamental research and learning-by-doing in the most promising (in terms of cost reductions) technologies across all possible carbon mitigation technologies, including CO_2 utilization. In other words, government expenditure needs to support innovation with positive spillovers rather than subsidize emissions reductions. The optimal budget depends on the size of the optimal stimulus needed to support innovation across a wide range of technologies. These guidelines are useful to set a benchmark for cost-effective regulation, but in many cases, regulators have to work with a predetermined budget to support specific technologies. The following sections address the best course of action in this case.

5.1.1.1 Working with Fixed Budgets

When a budget has been predetermined for interventions in a specific sector, economic theory suggests prioritizing policies and projects for funding using well-established rules for CBA. This requires estimating benefits and costs of all projects under examination and selecting the combination of projects that can be funded with the available budget and gives the largest total benefit (Boardman et al. 2017). Benefits and costs include monetized values of avoided negative externalities, positive knowledge spillovers, and all other positive market (e.g., labor) and nonmarket (e.g., health) impacts, during the entire lifetime of the project, discounted following standard rules. The benefit from reducing CO_2 emissions or storing carbon can be monetized using the Social Cost of Carbon (IWG 2021). Current regulation requires that CBA is done for all projects with spending estimated to be larger than $100M (Clinton 1993), but it can be beneficial for projects of smaller magnitude as well.

Limitations of CBA are well known and widely documented in the literature (Boardman et al. 2017). This includes the inability to validly monetize at least some nonmarket impacts and address normative considerations about the weight attributed to future generations and to different sectors of society. However, despite these limitations, CBA can provide useful guidance to constrain arbitrary allocation of limited budgets (Sunstein 2014, 2017).

5.1.1.2 Further Regulatory Needs

Although the major impediments to investment in CO_2 utilization technologies are the lack of regulation of CO_2 emissions and the lack of support for positive knowledge spillovers, other policy interventions are warranted for the following reasons.

5.1.1.2.1 Economies of Scale

Technologies and markets that rely on networks—such as pipelines to distribute CO_2 from capture to utilization sites—may exhibit economies of scale: As the investment in the network increases, the value of the network grows more than proportionally to its cost. These are called natural monopolies because it is optimal to concentrate all investments into one single project. For example, there is a clear economic case in favor of only one natural gas pipeline distribution network in cities. The standard prescription is to regulate the natural monopoly so that prices can provide fair profits for investors and fair costs for consumers, as in many privately owned electric power distribution networks. Alternatively, the government can retain ownership of the natural monopoly, as in the case of the federal interstate highway system. If technologies necessary for CO_2 utilization exhibit economies of scale, regulators will need to choose whether to own and operate the technology or to regulate its owners and operators.

5.1.1.2.2 Regulatory Uncertainty

Policy uncertainty—especially over the long term like the net-zero-emissions-by-2050 target—hinders investment and adoption of technologies for carbon capture, utilization, and storage (CCUS). This uncertainty may limit investments that would otherwise be profitable if investors perceived the government commitment to net-zero targets as fully credible. Public investment in infrastructure that supports CO_2 utilization may signal a policy commitment to create a market for low-carbon technologies, thus spurring private investment in related technologies. It is crucial here to understand that the problem stems from regulatory uncertainty rather than technological uncertainty. Public investments to reduce private risks, if not motivated by positive knowledge spillovers, are essentially subsidies, and the same caveats discussed above apply. Finally, regulators have to be careful to keep technological options as open as possible. One example might be using CBA to assess the benefits of flexible systems that can be adjusted progressively as new information becomes available.

5.1.1.2.3 Regulatory Costs

Regulation is needed to create market conditions for the diffusion of CO_2 utilization technologies and to account for other societal goals, such as safety measures. Regulators must carefully fill any regulatory gap that may induce uncertainty or lead to unintended consequences; at the same time, they need to be wary of imposing regulatory costs that do not generate clear societal benefits because such costs would unnecessarily constrain the CO_2 utilization economy. To provide insights on useful regulatory reform, Section 5.2 presents the current regulatory framework for CO_2 capture, transportation, utilization, and storage.

5.2 CURRENT REGULATORY FRAMEWORK FOR CARBON CAPTURE, UTILIZATION, AND STORAGE

5.2.1 Facilities Permitting

Permits are typically necessary for building and operating industrial facilities, for example, permits for construction or to discharge waste; permitting requirements for such facilities are generally well established. This is already true for carbon capture and hydrogen production facilities, both of which are envisioned to be used extensively as part of a carbon management strategy that includes CO_2 utilization. Table 5-1 was adapted from the White House Council on Environmental Quality (CEQ) Report to Congress on CCUS and a study examining the prospects for CCUS in the state of California (CEQ 2021; EFI 2020). The table outlines the general set of permits, authority, and agency or entity that has jurisdiction over issuance for a notional carbon capture facility (including one that is part of a retrofit to an existing industrial process) and a notional hydrogen production facility. Hydrogen production via water electrolysis and via methane reforming with carbon capture are included. Depending on the facility and industrial process(es) contained within it, multiple permits may need to be obtained.

TABLE 5-1 Authorizing Actions That May Be Necessary to Construct Facilities: Carbon Capture Including Retrofits and Hydrogen Production

Permit Name	Program or Authority	Description	Permitting Agency or Entity	Applicability[a]
Section 404 permit	Clean Water Act (CWA)	Under the CWA, a party must obtain a Section 404 permit from the U.S. Army Corps of Engineers (USACE) before discharging any dredged or fill material into waters of the United States. General 404 permits are issued to common activities that arise in projects. Otherwise, an Individual Permit is issued, which requires a more thorough process.	USACE	CC, WE, SMR + CCS
Federal Incidental Take permit	Endangered Species Act (ESA) § 10	If a species is listed in the state and federal ESA, consultation between the state-level cognizant entity and U.S. Fish & Wildlife Service (USFWS) is required to determine which agency is responsible for authorizing the incidental take.	USFWS	CC, WE, SMR + CCS
National Pollutant Discharge Elimination System (NPDES) permit	CWA	An NPDES permit is required if an entity discharges a pollutant from a point source to surface water. The state water control/quality authority, along with its nine subsidiary Regional Boards, issue NPDES permits. Many states have developed regulatory permit programs or reporting requirements or both for industries withdrawing greater than a specified threshold. The NPDES permit includes a detailed description of the plant providing basic information on all the sources of supply for the plant, the different ways in which water is used in the plant, and what water is included in the reported discharge values.	State entity	CC, WE, SMR + CCS
State-issued Incidental Take Permit	State-specific	A state-issued Incidental Take Permit authorizes the "take" of an endangered, threatened, or candidate species if the take is incidental to otherwise lawful activity, the impact of the authorized take is mitigated, and adequate funding is available to do so. Take, as defined by the ESA, refers to the harassment, harm, pursuit, hunting, shooting, wounding, killing, trapping, capture, or collection of the aforementioned species.	State entity	CC, WE, SMR + CCS
Prevention of Significant Deterioration (PSD) or New Source Review (NSR)	Clean Air Act (CAA) NSR	If a major stationary emission source is constructed or undergoes major modification, either an NSR or a PSD permit may be required prior to commencement of construction. If the source is located in an attainment area, PSD requires Best Available Control Technology to be determined from the source. If the source is in a nonattainment area, NSR requires Lowest Achievable Emissions Rate to be determined for the source. In addition, both PSD and NSR permitting processes require air quality analysis, additional impact analysis, and opportunities for public engagement. The NSR or PSD permit may require revision of a facility's Permit to Operate.	Appropriate EPA region	CC, WE,[b] SMR + CCS

TABLE 5-1 Continued

Permit Name	Program or Authority	Description	Permitting Agency or Entity	Applicability[a]
Joint environmental impact statement (EIS)/ environmental impact report (EIR)	State-specific and National Environmental Policy Act (NEPA)	When a project requires federal and state approvals, a joint EIS/EIR may be required (or Finding of No Significant Impact/Negative Declaration if there is no significant environmental impact). In this case, one state and one or more federal agency cooperate to reduce the duplication of any processes. In some cases, due to the divergence in expectations, the lead agency for the state-entity assessment may determine that an EIR is necessary, while the NEPA lead agency decides that there are no potential significant environmental impacts. When that happens, the agencies write a joint environmental assessment EIR with an explanation of why the federal agency determined no potential significant environmental impacts.	Situational (one state and one federal agency)	CC, WE, SMR + CCS
Local Conditional Use Permit (CUP)	State-specific	A CUP allows a city or county to consider special use of its land that may be favorable to the community but is not allowed within a zoning district. The project or development is proposed in a public hearing and if it is approved, it allows flexibility within the zoning ordinance with stipulations. The CUP may be subject to a state-level environmental quality assessment, which may lead to an EIR before a public hearing can occur. The project must also fit in the context of a city's or county's general plan, which lays out the long-term plan for the community. Sometimes, developers will apply for a general plan amendment instead of a CUP.	Cities or counties	CC, WE, SMR + CCS
Fire safety permit	State-specific	A fire safety permit is required of all new and retrofitted facilities to ensure compliance with various federal, state, and local standards, especially those related to combustion and explosive processes and products.	Cities, counties, or state	CC, WE, SMR + CCS
Utility approvals	Service territory and/or state specific	Utility approvals include those for water, electricity, and (if any is used) natural gas. Utility interconnection requirements may or may not include interconnection studies, depending on the size and configuration of the facility and the topography of the services required.	Utility, cities, counties, or state	CC, WE, SMR + CCS

[a] CC = carbon capture, WE = water electrolysis, SMR + CCS = steam methane reforming with carbon capture and storage.

[b] Where pure/filtered water is used within the electrolysis process to produce hydrogen, the only by-product is O_2. In this case, no criteria pollutants are produced, and therefore there is no requirement to obtain air permits.

SOURCE: Adapted from Council on Environmental Quality, 2021, *Report to Congress on Carbon Capture, Utilization, and Sequestration*, Washington, DC, https://www.whitehouse.gov/wp-content/uploads/2021/06/CEQ-CCUS-Permitting-Report.pdf; and Energy Futures Initiative, 2020, "An Action Plan for Carbon Capture and Storage in California: Opportunities, Challenges, and Solutions," https://energyfuturesinitiative.org/reports/an-action-plan-for-carbon-capture-and-storage-in-california.

One regulatory gap focused on hydrogen relates to standards associated with on-site storage. The National Fire Protection Association Rule 2: Hydrogen Technologies Code provides standards for setbacks and protections for storage facilities containing up to 175,000 gallons. However, currently there are hydrogen storage facilities in the United States that far exceed this volume, whereby the developer establishes its own setback and safe zoning criteria.

5.2.2 Pipeline Permitting

Permitting and compliance requirements for pipelines are well established (see Table 5-2). Pipelines have been built, utilized, and maintained to transport materials for over a century. While large-scale transportation of CO_2 is somewhat novel, pipeline transport of CO_2 has occurred safely for approximately 50 years. In fact, according to recent analytical work (Hawkins et al. 2021), CO_2 pipelines were determined to be among the safest in the industry. Despite their record of safety, no technology is perfect, and CO_2 pipelines would benefit from increased rigor in their compliance, inspection, and enforcement programs (PHMSA 2019). The Pipeline and Hazardous Materials Safety Agency (PHMSA) currently has no regulations applicable to pipelines transporting CO_2 as a gas, liquid, or in a supercritical state at concentrations of CO_2 less than 90 percent. However, they do have the authority to make such regulations. PHMSA recently published its intent to develop new safety measures to strengthen oversight of CO_2 pipelines (PHMSA 2022), and the U.S. Department of Energy (DOE) plans to incorporate these measures in their CO_2 pipeline research, development, and demonstration (RD&D) projects (DOE 2022). Added safety measures would provide even greater safety for those that may be impacted by pipeline incidents.

5.2.3 Regulatory Frictions or Uncertainties

Permitting a CCUS project is similar to the process for permitting any industrial activity. The pathway for regulating CCUS projects is established, and the precise mixture of permits, reviews, and approvals needed for a particular project will be determined by the specific details of the project (CEQ 2021). Indeed, some combinations of capture and utilization processes may have relatively little permitting friction beyond that of any other kind of industrial facility. This is especially true of relatively small-scale, discrete processes involving new or modified industrial facilities, for example, a slipstream capture of a small portion of the pure CO_2 emitted from an ethanol plant, transported by truck, and utilized as part of a concrete curing process. The cost of capture, coupled with incentives to geologically sequester captured CO_2, may induce consideration of larger capture facilities, where only a portion of CO_2 captured is being diverted to utilization. In this case, where CO_2 utilization is only a sub-process within a larger CCUS project, regulatory frictions may arise. It is well documented that full-value-chain CCUS, including transportation, suffers from long lead times due to the integrated nature of the project and the fact that it covers multiple regulatory boundaries across various authorities having jurisdiction (CEQ 2021; LEP 2021). Often, there is no lead agency charged with "owning" the process, coordinating the basket of permits and approvals in collaboration with the project sponsor. The lead agency would normally be the agency with general governmental powers, such as a city or county, rather than an agency with a single or limited purpose (e.g., state agency). Lead agencies also can be a district that will provide a public service or public utility.

Projects that trigger a federal nexus (e.g., utilize federal lands or significant federal funds) may have additional permitting complexities. In 2022, CEQ recognized the permitting complexities surrounding the end-to-end carbon capture value chain, especially given the relative nascence of CCUS as a commercial endeavor. CEQ produced guidance to facilitate reviews associated with the deployment of CCUS and to promote the efficient, orderly, and responsible development and permitting of CCUS projects at an increased scale in line with the Biden administration's climate, economic, and public health goals (CEQ 2022). The vast majority of this guidance applies to CCU projects alone and aims at establishing clear, consistent rules within agencies to reduce coordination frictions, given that CCUS represents a new form of industrial organization that current agencies have to accommodate. Guidance includes directing agencies to consider developing programmatic environmental reviews to increase the efficiency and effectiveness of the CCUS permitting process and developing interagency memoranda of understanding to facilitate collaboration on CCUS activities, as well as CEQ facilitation of interagency collaboration between the

TABLE 5-2 Authorizing Actions That May Be Necessary to Construct Pipelines

Agency	Nature of Authorizing Action	Authority
Federal permits, approvals, and reviews		
Bureau of Land Management	Grants rights-of-way and issues temporary use permits	Section 28 of the Mineral Leasing Act of 1920
	Issues materials sales contracts	Materials Act of 1947, as amended; 30 U.S.C. §§ 601, 602; 43 CFR Part 3600
	Issues antiquities and cultural resource use permit to excavate or remove cultural resources on federal lands	Antiquities Act of 1906, 16 U.S.C. §§ 431–433 Archaeological Resources Public Protection Act of 1979, 16 U.S.C. §§ 470aa–470mm; 43 CFR Part 3
	Approves herbicide use on federal lands	BLM Manual 9011.1, Guidelines for Conducting Chemical Pest Control Program
U.S. Fish & Wildlife Service	Section 7 consultation process for endangered or threatened species	Endangered Species Act of 1973, 16 U.S.C. §§ 1531 et seq.
U.S. Forest Service	Grants special use authorizations for rights-of-way	30 CFR § 251.53
Federal Highway Administration	Issues permits to cross federal-aid highways	23 U.S.C. §§ 116, 123 23 CFR Part 645, Subpart B
Pipeline and Hazardous Materials Safety Administration	Regulates safe operation of CO_2 pipelines and Regulates safe operation of hydrogen pipelines	49 CFR Part 195 and 49 CFR Part 192
U.S. Army Corps of Engineers	Issues Section 404 permit (nationwide) for placement of dredged or filled material in waters of the United States	Section 404 of the Clean Water Act of 1972 (40 CFR Parts 122–123); 33 U.S.C. § 1344; 33 CFR Parts 323, 325
Bureau of Alcohol, Tobacco, and Firearms	Issues permits to purchase, store, and use explosives	Section 1102(a) of the Organized Crime Control Act of 1970, 18 U.S.C. §§ 841–848; 27 CFR Part 181
Advisory Council on Historic Preservation	Performs review and compliance activities related to cultural resources	Section 106 of the National Historic Preservation Act, 16 U.S.C. § 470; 36 CFR Part 80
State and local authorizations		
Department of Environmental Quality, Water Quality Division	Issues National Pollution Discharge Elimination System permit for discharges; approves Stormwater pollution prevention plan	State environmental quality statute
	401 Water Quality Certification	Section 401 of the Clean Water Act
Highway department	Issues permits for oversize and overweight loads	State transportation department
	Issues encroachment permits for state highways	State transportation department
State land board	Issues easements to cross state lands	State land board statute
State engineer's office	Grants permit to appropriate water for hydrostatic testing, dust control, and other uses	State engineer statute
State historic preservation office	Reviews compliance activities related to cultural resources	Section 106 of the National Historic Preservation Act, 16 U.S.C. § 470; 36 CFR Part 80
County commissioners	Issues road crossing permits, land-use permits, and licenses	County zoning regulations
County health departments	Permits temporary sanitation facilities	County sanitation regulations

various federal and state agencies involved in the pipeline regulatory process. While these recommendations could reduce the permitting time and complexity, to what extent they will do so is unclear. Often, processes like these improve iteratively through successive applications by sponsors seeking to develop projects. The Infrastructure Investment and Jobs Act (IIJA, Public Law 117-58) includes authorization and funding that expands an existing DOE program for carbon storage validation and testing to include commercialization and associated CO_2 transport infrastructure. The commercialization program is intended to facilitate new or expanded CCS and associated CO_2 transport infrastructure and includes funding for feasibility, site characterization, permitting, and construction, giving priority to those storing substantial amounts of CO_2 or those collecting from multiple capture facilities (IIJA § 40305).

5.2.4 Economic Policy Friction or Uncertainties

The signing into law of the Inflation Reduction Act (IRA) in August 2022 significantly enhanced the magnitude of the Section 45Q Carbon Capture Credit. Notably, provided a project adheres to prevailing wage and apprenticeship requirements, the tax credit has been increased from $35/tonne to $60/tonne for utilization. This value can effectively cover the cost of capture and transportation of CO_2 applied to such point sources as gas processing facilities and ethanol production in a generic project setting. Moreover, the IRA lowered the threshold of eligibility for a facility to qualify for the tax credits, namely 12,500 tonnes of qualified CO_2 during the taxable year for non-electricity generating facilities and 18,750 tonnes for electricity generating facilities. These values are important for CO_2 utilization projects, since initially the amount utilized presumably would be relatively small, especially as a stand-alone project not connected to a large injection (sequestration) project. Finally, the IRA allows for tax credit transferability and refundability provisions, thereby increasing not only the number and type of potential investors in carbon utilization (CU) projects, but also increasing the value of the tax credits because credit monetization in some cases may largely avoid the use of complex tax equity deals. Taken together, the IRA has materially reduced the cost barrier to some CU processes. Coupled with the $310 million made available through Section 40302 of the IIJA for carbon utilization market development (a grant program for state and local governments to procure and use products derived from captured carbon oxides), CU product economics will benefit from these supply-side and demand-side provisions.

As it stands today, the 45Q tax credit for CO_2 utilization allows for any commercial market (new or existing) that can demonstrate a net reduction in CO_2e (carbon dioxide equivalent) without restriction, net reduction sufficing in the short term to support early development and knowledge spillover on the way to achieving net-zero. While some may have preferred a narrower definition of commercial markets for CO_2, it is important to recognize the interaction that carbon capture, utilization, and storage incentives such as 45Q and the California Low Carbon Fuel Standard have with traditional markets for CO_2, such as food and beverage, the third largest market for CO_2 in the United States today. Allowing CO_2 emitters that demonstrate a net reduction in CO_2e, such as ethanol plants, to qualify for 45Q enables those lower CO_2e sources to be used by existing markets, like food and beverage companies. These markets would otherwise continue to obtain CO_2 from fossil or other higher-CO_2e sources, undermining the net CO_2e reduction goal of 45Q. Notwithstanding the previous argument, it is critical to note that the structure of the 45Q tax credit incentivizes geologic storage over utilization ($85/tonne versus $60/tonne), which may encourage facilities located near favorable geologic formations to sequester CO_2 underground as opposed to supplying industry.

Beyond the magnitude of the 45Q tax credit is its duration, currently set at 12 years. Typically, industrial facilities are long-lived assets designed for at least 25 years, with depreciation and business plans scaled accordingly. While it is possible to develop investment cases with attractive returns under a 12-year regime, those facilities would cease to operate in year 13 if they are unlikely to yield sufficient revenue without the 45Q credit. Essentially, the duration of the 45Q tax credit and the activities that it intends to induce may be mismatched, thereby causing an underinvestment in viable CO_2 utilization projects. This case may be obviated if there is a sufficient price on carbon or some other support mechanism present at the time of potential closure, but this future uncertainty acts as an inhibitor to investment. It is important to continue studying the market impacts of and obtain input from industry on different incentives for CO_2 utilization.

Relatedly, the IRA established a 45V Clean Hydrogen production tax credit (PTC), eligible for direct pay, valued at $3/kg if the life cycle GHG emissions rate is less than 0.45 kg CO_2e per kilogram of hydrogen, and adjusted proportionately downward for greater life cycle GHG emission rates. The 45V credit may have a material effect on the cost competitiveness of products that use clean hydrogen as an input to CO_2 utilization. One potential friction with the current 45Q and 45V tax credits is their exclusivity: A single taxpayer cannot benefit from both the clean hydrogen PTC and the 45Q credit, thereby requiring joint ventures or other project structuring to attempt to benefit from both, which adds a layer of complexity.

Specific to CO_2 transportation, the IIJA authorizes and appropriates $2.1 billion for grants and loans to build CO_2 common carrier infrastructure for eligible projects expected to cost $100 million or more. Loans aim to help eligible projects attract investment and begin earlier than would otherwise be possible. Grants are targeted to cover the cost of constructing a facility capable of accommodating future growth due to demand for CO_2 transport. The law calls for the Secretary of Energy to prioritize projects that are large-capacity, have common-carrier infrastructure, have demonstrated demand for infrastructure from CO_2 capture facilities, represent geographic diversity, and have site infrastructure within existing corridors to minimize environmental disturbance and other siting concerns (IIJA 2021, § 40305).

The cost of input electricity creates an additional challenge for developers of CO_2 utilization technologies. Electricity is a critical component for many CO_2 utilization processes and, in a net-zero future, must be carbon-emissions free, either directly or indirectly through a power purchase agreement. Electricity unit cost might be relatively high if facilities making use of the captured CO_2 (or producing input hydrogen) do not have access to wholesale electricity prices. This may be the case for facilities located outside a deregulated electricity market and/or those required to purchase electricity at retail rates from a utility or electricity retailer due to projected small demand. In this case, unless there is a low-cost industrial rate, or a specially designed rate (TPU 2021) that accommodates the operational characteristics of small consumers—for example, small electrofuels producers—then the relative cost of the product based on captured CO_2 would be affected adversely.

A final economic policy friction relates to the interstate movement of hydrogen through pipelines. The Federal Energy Regulatory Commission (FERC) regulates interstate commerce related to natural gas, electricity, and oil. FERC has broad authority under the Natural Gas Act (NGA 2006) to regulate the construction of interstate natural gas pipelines, as well as the rates and tariffs governing the interstate transportation of natural gas. However, hydrogen is not natural gas or an artificial gas, and therefore the NGA is silent on hydrogen and does not grant FERC jurisdiction to regulate interstate pipelines transporting pure hydrogen. As an aside, FERC may have jurisdiction to regulate hydrogen introduced into interstate natural gas pipelines to supplement or displace natural gas. Until this question is resolved, perhaps through an amendment to the NGA, then project sponsors will be hesitant to plan for such pipelines, which could work against the cost competitiveness of products attempting to use hydrogen as an input to CO_2 utilization processes unless hydrogen production can occur on demand on-site.

5.3 SOCIETAL ACCEPTANCE AND ENVIRONMENTAL JUSTICE

Widespread deployment of carbon capture and utilization will lead to diverse environmental, economic, and societal impacts. The application of CBA, as recommended in Section 5.1, ensures that all these impacts are considered and, whenever possible, monetized. However, the recommendation to select projects to maximize societal net benefits does not take into consideration the distribution of costs and benefits among different groups in society. Although quantifying the distributional effects of a project should be part of CBA, there are no objective rules to rank projects based on their distributional effects (Boardman et al. 2017). To assess equity issues, policy makers have to use additional normative criteria that reflect society's views and preferences about justice.

5.3.1 Distributional Effects

Impacts of CO_2 utilization projects can vary among regions, demographic groups, and communities. Even if society as a whole will benefit, some groups—for example, those residing near production facilities—may be negatively affected. These distributional impacts can be complex. For example, consider the case of mitigating CO_2

emissions from vehicles via either converting the fleet to electric vehicles and decarbonizing the electric system, or retaining an internal combustion engine fleet and producing synthetic net-zero carbon fuels for use in internal combustion engine vehicles. Both technologies could have the same net reduction in global CO_2 emissions, but the impact on local pollution is likely to be different across communities, within the country, and internationally. Carbon capture and fuel production may have local pollution impacts. Using the fleet of internal combustion engine vehicles with synthetic fuel will produce health-harming criteria pollutants from vehicle tailpipes, while electric vehicle fleets will not emit pollution at the tailpipe. Electric vehicles have higher manufacturing energy requirements than internal combustion engines with potentially negative impacts in terms of pollution and resource use. Extraction of minerals used for batteries is highly polluting. Because low-income households and historically disadvantaged groups live in areas more environmentally fragile and more exposed to pollution, the local impact of either technology may disproportionately fall on groups that already suffer disproportionately from pollution, despite benefits experienced globally. The different nature of pollution from the two technology solutions will generate different impacts within these very same disadvantaged communities that need further consideration.

Regulators, in cooperation with affected communities and CO_2 utilization project developers, need to assess these distributional impacts, including consideration of equity and justice for historically disadvantaged groups. One way to address disproportionate impacts is for the project beneficiaries to provide compensation to those who are affected negatively. Compensation can facilitate acceptability of the project but also may be rejected by the parties involved. Based on careful assessment of benefits, costs, distributional effects, and potential compensatory measures, regulators might choose not to invest in projects that would generate a net benefit for society but have unavoidable and unacceptable equity implications.

5.3.2 Environmental Justice

The environmental justice movement aims to ensure that all environmental benefits and costs are shared equally and that historical damages to disadvantaged communities are addressed. This movement intersected the civil rights movement (IEP n.d.); Title VI of the Civil Rights Act of 1964 (42 U.S.C. § 2000d et seq.) enabled disadvantaged communities to sue on the basis of environmental discrimination, as Title VI prohibits discrimination on the basis of race, color, or national origin in any program or activity that receives federal funds or other federal financial assistance. In subsequent decades, the disproportionate environmental impacts felt by underrepresented communities continued to be exposed, and federal environmental legislation advanced, including passage of the National Environmental Policy Act (NEPA, 42 U.S.C. §§ 4321–4370h) and the Clean Air Act (CAA, 42 U.S.C. § 7401 et seq.). In 1991, the 17 Principles of Environmental Justice were adopted at the First National People of Color Environmental Leadership Summit as a foundation for grassroots efforts (Principles of Environmental Justice 1996).

Prompted by advocates of justice, federal and state governments have worked to implement processes to address environmental justice in laws, regulations, and policies (see, e.g., Biden 2021; Clinton 1994). Recent federal efforts to advance environmental justice include the Justice40 Initiative (Justice40 2021), which was established in 2021 and codified in Executive Order 14008 (Biden 2021). The U.S. Department of Energy's implementation of Justice40 focuses on improving parity and opportunities for disadvantaged communities, including decreasing energy and environmental burdens and increasing access and adoption of clean energy technologies (DOE-OEID n.d.). To improve disadvantaged communities' opportunities for clean energy, DOE prioritizes access to low-cost capital, increased enterprise creation and contracting, jobs and training, energy resiliency, and energy democracy. Box 5-1 outlines concepts of environmental justice.

Environmental justice comes into play for CO_2 utilization both in siting of potentially polluting industrial processes in and around communities, and in including CO_2 utilization as a part of climate change mitigation, in which CO_2 utilization technologies may be in competition with other options for reducing climate pollutants. From an equity perspective, many environmental justice organizations view CO_2 utilization as a means of perpetuating industries that have caused and continue to harm disadvantaged communities, and they have significant concerns about the techno-economic viability and safety of the attendant infrastructure (see, e.g., Amsalem and Bogdan Tejeda 2022; Climate Justice Alliance 2022; Flores-Jones 2022). Specific industrial or power plant facilities in impacted communities, which might otherwise be phased out of use under a net-zero carbon emissions scenario that limits or does not allow CO_2 utilization, may continue operation and production of the attendant pollution. Such facilities even may expand operations, given the increased need for energy to capture CO_2. Additional infrastructure needs, such as pipeline siting or development of hubs around existing industrial facilities, similarly are a health and operational safety concern to surrounding neighborhoods. Specific impacts of different facets of CO_2 utilization technologies and infrastructure can be found in Chapter 4.

Environmental justice is about equitably apportioning both risks and benefits. Carbon180, a nonprofit organization focused on equitable carbon removal solutions, lays out guiding principles for just application of carbon removal (Kosar and Suarez 2021), which are also relevant to CO_2 utilization technologies:

- The benefits of carbon removal solutions must be equitably distributed.
- Public engagement must be robust and involve seeking input from groups throughout the development and deployment of carbon removal solutions.
- Safeguards are needed to ensure that adverse impacts are not borne by disadvantaged communities.
- The socioeconomic consequences and distributional impacts of carbon removal solutions need to be evaluated alongside their technological and economic attributes.
- Carbon removal is seeking to address a challenge that is both local and global, and therefore should incorporate justice across temporal and spatial scales.

Justice considerations for CO_2 utilization infrastructure differ from those of carbon removal in favorable and unfavorable ways. While both utilization and removal solutions can mitigate climate change and locally reduce criteria pollutants, CO_2 utilization differs in that it additionally results in goods with monetary and other value, which can provide advantages to potential host communities of utilization infrastructure over that of carbon removal infrastructure. On the other hand, CO_2 utilization can have disadvantages in comparison to carbon removal, such as a tendency to co-locate with existing industrial facilities, likely in communities already bearing adverse impacts. In both cases, it is important to implement a justice-oriented process when siting and developing this infrastructure. The next section highlights current approaches for equitable community engagement.

5.3.3 Current Approaches to Communication and Productive Community Engagement in Planning

Current public engagement standards and guidance for just implementation of industrial development have been described as insufficient and lacking in rigor to address current challenges (Kosar and Suarez 2021). The core issue is the existence of a power imbalance between developers/operators and impacted communities, particularly communities composed of historically underrepresented groups. Local and national government officials, as well as scientists and nongovernmental organizations, have varying levels of agency and economic and social stake in the outcome. Implementing a process that will allow underrepresented communities to have their needs fairly considered is important in light of this imbalance.

Early community engagement is critical to any large-scale infrastructure project. Absent community engagement, a project almost certainly will fail, encounter delays, or require expensive reworking. Community engagement could mean a project will not move ahead in its originally proposed form but could establish the viability of a project, as well as avoid false starts and costly adjustments down the line. For this process also to be just, each community would have its needs given equal consideration regardless of relative representation and power. Project developers likely will see the most success if they are honest about the impacts, site projects in communities that want them, and show or share the benefits with the impacted community (Nielson et al. 2022). Individual community engagement activities can take different forms depending on the goal (see Table 5-3), for example, whether to inform or receive direct feedback, and might vary throughout the project planning and development process. Communities also see value in sustained engagement, ensuring that facilities or infrastructure in their neighborhoods achieve and maintain the agreed-upon, long-term outcomes (Romero-Lankao 2022).

Numerous governmental and nongovernmental organizations have developed frameworks for community engagement (Brooks 2022; CEQ 2021, 2022; Cochran and Denholm 2021; EPA 2021b; Forbes et al. 2008; Kosar

> **BOX 5-1**
> **Concepts in Justice and Equity**
>
> The concepts of justice and particularly environmental justice, equity, disadvantaged communities, and historically underrepresented groups are all important to analyzing, understanding, and addressing inequitable distribution of the benefits and costs of CO_2 utilization technologies.
>
> - Justice
>   - Social arrangements that permit all (adult) members of society to interact with one another as peers (Nancy Fraser, quoted in Cochran and Denholm 2021).
>   - Procedural justice—fairness in decision-making processes (Kosar and Suarez 2021); the ability of people to be involved in decision-making processes around energy-system infrastructures and technologies (Cochran and Denholm 2021).
>   - Distributive justice—equitable allocation of resources, risks, impacts, and benefits across society (Kosar and Suarez 2021); distribution of benefits and burdens across populations (Cochran and Denholm 2021).
>   - Reparative justice—repairing previous harms committed through violations and crimes (Kosar and Suarez 2021).
>   - Transformative justice—spurring changes in society's current systems and structures (Kosar and Suarez 2021).
>   - Recognition justice—understanding the historical and present bias for social inequalities and the acknowledgment or dismissal of marginalized and deprived communities in relation to energy systems (Cochran and Denholm 2021).
> - Equity
>   - Achieved results where advantage and disadvantage are not distributed on the basis of social identities (Initiative for Energy Justice 2019).
> - Environmental justice
>   - The fair treatment and meaningful involvement of all people regardless of race, color, national origin, or income with respect to the development, implementation, and enforcement of environmental laws, regulations, and policies. Fair treatment means that no population, due to policy or economic disempowerment, is forced to bear a disproportionate burden of the negative human health or environmental impacts of pollution or other environmental consequences

TABLE 5-3 Participatory Governance Styles: Los Angeles Department of Water and Power and Other Examples

Participatory Governance Style	City of Los Angeles, LADWP, and Other Examples
Educative forum	Community meetings and presentations Community assemblies through the Office of Climate Emergency and Mobilization Deliberative polling
Participatory advisory panel	LA100 Advisory Group, Oregon Health Decisions, Citizen Summit
Participatory problem-solving collaboration	Community Partnership Grants Program, Neighborhood Councils, Citizen Summit, Neighborhood Planning Initiative
Participatory democratic governance	Participatory budgeting

SOURCE: Adapted from J. Cochran and P. Denholm, eds., 2021, *LA100: The Los Angeles 100% Renewable Energy Study,* NREL/TP-6A20-79444, Golden, CO: National Renewable Energy Laboratory, https://maps.nrel.gov/la100.

resulting from industrial, municipal, and commercial operations or the execution of federal, state, local, and tribal programs and policies. Meaningful involvement means people have an opportunity to participate in decisions about activities that may affect their environment and/or health; the public's contribution can influence the regulatory agency's decision; community concerns will be considered in the decision-making process; and decision makers will seek out and facilitate the involvement of those potentially affected (EPA 2021a).

- Disadvantaged community
  - A community that suffers the most from a combination of health, economic, and environmental burdens. These burdens include high unemployment, air and water pollution, and poverty (Kosar and Suarez 2021).
  - Socially disadvantaged—those who have been subjected to racial or ethnic prejudice or cultural bias based solely on their identities as members of groups without regard for individual qualities (Kosar and Suarez 2021).
  - Economically disadvantaged—those whose ability to access credit opportunities and generate capital has been impaired, negatively impacting their standard of living (Kosar and Suarez 2021).
- Historically underrepresented (modified from Emory University n.d.)
  - This term refers to groups who have been denied access and/or have suffered past institutional discrimination in the United States and, according to the Census and other federal measuring tools, includes African Americans, Asian Americans, Hispanics or Chicanos/Latinos, and Native Americans.
  - This is revealed by an imbalance in the representation of different groups in common pursuits such as education, jobs, and housing, resulting in marginalization for some groups and individuals and not for others, relative to the number of individuals who are members of the population involved. Other groups in the United States have been marginalized and are currently underrepresented. These groups may include but are not limited to:
    - Other ethnicities
    - Veterans
    - People with disabilities
    - Lesbian, gay, bisexual, and transgender individuals
    - Different religious groups, and
    - People from different economic backgrounds

and Suarez 2021; Tollefson et al. 2017). Common among these approaches is the need to involve communities early, evaluate and provide information on potential impacts to the affected communities, establish trust, and allow community engagement to materially change the outcome of the proposed project. An example of such community engagement that has been used predominantly in siting of nuclear waste storage facilities is termed consent-based siting (DOE-NE n.d.).

Consent-based siting is an approach to siting facilities that focuses on the needs and concerns of people and communities. Communities participate in the siting process by working carefully through a series of phases and steps with the Department (as the implementing organization). Each step and phase helps a community determine whether and how hosting a facility to manage spent nuclear fuel is aligned to the community's goals. By its nature, a consent-based siting process must be flexible, adaptive, and responsive to community concerns. Thus, the phases and steps are intended to serve as a guide, not a prescriptive set of instructions. Working through the consent-based siting process collaboratively builds a mutual trust relationship between DOE and a potential host community. Potential outcomes from the consent-based siting process could include either a negotiated consent agreement or a determination that after exploring the option in good faith, the community is not, in fact, interested in serving as a host. Both are successful outcomes.

Another example of community engagement requirements for siting industrial facilities is the New Jersey Department of Environmental Protection (NJDEP) Environmental Justice Rule Proposal, released in June 2022. It lays out the process by which permit applicants seeking to renew, open, or expand pollution-generating facilities in overburdened communities assess the facility's potential impact and engage with the community. Applicants must complete an Environmental Justice Impact Statement and demonstrate that they either avoid disproportionate impact on overburdened communities or serve a compelling public interest in said communities while still minimizing impact. For direct community engagement, applicants are required to hold a public hearing and respond to comments, which will be considered by NJDEP when determining whether to authorize the applicant to proceed (NJDEP 2022).

This committee will address environmental justice approaches further in its subsequent report.

5.4 FINDINGS AND RECOMMENDATIONS FOR POLICY, REGULATORY, AND SOCIETAL CONSIDERATIONS FOR CO_2 UTILIZATION

FINDING 5.1 Economic Tools to Support CO_2 Utilization. The most cost-effective way to promote the diffusion of CO_2 utilization technologies is to internalize the carbon externality (e.g., with a carbon tax or emissions trading scheme) and to subsidize knowledge creation in CO_2 utilization technologies. For knowledge creation, grants can promote fundamental research (learning by researching), while tax credits and procurement subsidies can stimulate incremental knowledge generation from the construction and operation of pilot plants and demonstration units (learning by doing).

FINDING 5.2 Limitations of Subsidies. Policies that subsidize the adoption of specific mitigation technologies—including policies related to carbon capture, utilization, and storage—can lead to excessive use of the subsidized technologies if not accompanied by a hard limit based on level of technology diffusion. Without such a limit, a subsidy such as tax rebates (e.g., tax code 45Q) can create perverse incentives to continue operating inefficient technologies with high emissions, or even the creation of emissions that would not otherwise exist. Similarly, credits to capture CO_2 for enhanced oil recovery can result in a subsidy for fossil-fuel production.

FINDING 5.3 Impacts of Policy Uncertainty. Investment and adoption of technologies for CO_2 capture, utilization, and storage are hindered by policy uncertainty, especially over the long term, as in the global target of net-zero emissions by 2050. This uncertainty may limit investments that would otherwise be profitable if investors perceived the government commitment to net-zero targets as fully credible. Public investment in infrastructure that supports CO_2 utilization may signal a policy commitment to create a market for low-carbon technologies, thus spurring private investment in related technologies.

> **RECOMMENDATION 5.1.** Efficient regulation of CO_2 emissions by the U.S. Environmental Protection Agency and state agencies should rely on a mixture of policy tools that uniformly and credibly penalize all greenhouse gas emissions across all sources over the entire policy horizon (such as a CO_2 emissions tax or a carbon trading scheme) and a mixture of policy tools that subsidize knowledge creation at all stages (such as research investments). Subsidies for emissions reduction technologies (perhaps partially offset from revenues collected via a carbon penalty) may help create learning externalities and may signal a strong policy commitment for the long term, but they should be used carefully to avoid perverse incentives from excessive investment in the targeted technologies (i.e., avoid subsidies that encourage continued emissions where they would otherwise be eliminated and avoid incentivizing new negative externalities).

FINDING 5.4 Regulatory Considerations. Complex regulatory frameworks, which are necessary to define markets, protect public safety, and achieve societal goals such as environmental justice, may generate costs that reduce or potentially even prevent investment in infrastructure supporting CO_2 utilization markets and can slow down the diffusion of CO_2 utilization technologies needed to support a net-zero future.

RECOMMENDATION 5.2. All states should craft regulation that is efficient and clearly communicated to achieve public policy goals while providing a usable framework for participation in CO_2 utilization markets without unnecessarily penalizing the deployment of CO_2 utilization projects across the value chain.

RECOMMENDATION 5.3. The U.S. Department of Energy should develop a CarbonStar program that labels products based on their carbon intensity to create transparency for buyers.

FINDING 5.5 Permitting Landscape. The many permits necessary for CO_2 utilization projects typically are processed through multiple federal, state, and local agencies, and each individual permit can require involved analyses.

RECOMMENDATION 5.4. A single agency or entity should be appointed to coordinate the permitting and authorization process for CO_2 utilization projects, guiding developers through the process of dealing with the multiple states and localities to obtain the required permits.

FINDING 5.6 Regulations for CO_2 Transportation. The Pipeline and Hazardous Materials Safety Administration currently has no regulations applicable to pipelines transporting CO_2 as a gas, liquid, or in a supercritical state at concentrations of CO_2 less than 90 percent. However, they do have the authority to make such regulations.

FINDING 5.7 Regulations for Hydrogen Transportation. The Federal Energy Regulatory Commission does not have authority under the Natural Gas Act to regulate pure hydrogen across state boundaries.

RECOMMENDATION 5.5. Congress should require the U.S. Department of Energy and the U.S. Department of Transportation to research and report on the full spectrum of regulations required to site, develop, and operate large-scale, interstate hydrogen infrastructure for transportation and storage.

FINDING 5.8 Cost-Benefit Analysis. Widespread deployment of CO_2 utilization may lead to diverse environmental, economic, and societal impacts. Cost-benefit analysis applied to a CO_2 utilization project, as to other government projects and policies, can provide the appropriate framework for dealing with a multidimensional problem, choosing how to invest scarce public resources to maximize aggregate total societal benefits, including nonmarket impacts such as pollution damages and employment benefits. Cost-benefit analysis rules can be used to estimate distributional impacts, but cannot be used to judge if an action is fair or not. On the basis of careful assessment of benefits, costs, distributional effects, and potential compensatory measures, regulators may choose not to invest in projects that would generate a net benefit for society but have equity implications that are deemed unavoidable and unacceptable.

FINDING 5.9 Community Engagement. Disadvantaged communities have not had substantive agency in affecting development of infrastructure that often negatively impacts them. Community engagement is a process that can enable just and equitable outcomes for those populations. Early and ongoing community engagement is important for a project's ability to move forward with community support. Absent such early and ongoing community engagement, infrastructure projects are likely to fail, encounter delays, or require expensive reworking.

RECOMMENDATION 5.6. Regulatory authorities in charge of siting infrastructure should account for distributional impacts of CO_2 utilization projects through a process that considers equity and justice for disadvantaged groups, engages impacted communities early and throughout the project planning, and allows for alteration of project design and implementation.

5.5 REFERENCES

Aghion, P., A. Dechezleprêtre, D. Hémous, R. Martin, and J. Van Reenen. 2016. "Carbon Taxes, Path Dependency, and Directed Technical Change: Evidence from the Auto Industry." *Journal of Political Economy* 124(1):1–51. https://doi.org/10.1086/684581.

Amsalem, G., and V. Bogdan Tejeda. 2022. "EPA Urged to Reject Carbon Capture Projects in Central California." Center for Biological Diversity. https://biologicaldiversity.org/w/news/press-releases/epa-urged-to-reject-carbon-capture-projects-in-central-california-2022-06-29.

Arrow, K.J. 1962. "Economic Welfare and the Allocation of Resources for Invention." Pp. 609–626 in *The Rate and Direction of Inventive Activity*. Princeton, NJ: Princeton University Press. https://doi.org/10.1515/9781400879762-024.

Biden, J.R. 2021. Executive Order 14008: "Tackling the Climate Crisis at Home and Abroad." *Federal Register* 86(19):7619–7633. https://www.govinfo.gov/content/pkg/FR-2021-02-01/pdf/2021-02177.pdf.

Blanco, G., H.C. de Coninck, L. Agbemabiese, E.H.M. Diagne, L.D. Anadon, Y.S. Lim, W.A. Pengue, et al. 2022. "Innovation, Technology Development and Transfer." Chapter 16 in *Climate Change 2022: Mitigation of Climate Change*. Contribution of Working Group III to the Sixth Assessment Report of the Intergovernmental Panel on Climate Change. Cambridge, UK: Cambridge University Press. https://www.ipcc.ch/report/ar6/wg3/downloads/report/IPCC_AR6_WGIII_Chapter_16.pdf.

Boardman, A.E., D.H. Greenberg, A.R. Vining, and D.L. Weimer. 2017. *Cost-Benefit Analysis: Concepts and Practice*. Cambridge, UK: Cambridge University Press.

Bovenberg, A.L. 1999. "Green Tax Reforms and the Double Dividend: An Updated Reader's Guide." *International Tax and Public Finance* 6(3):421–443. https://doi.org/10.1023/A:1008715920337.

Bovenberg, A.L., and L. Goulder. 1994. *Optimal Environmental Taxation in the Presence of Other Taxes: General Equilibrium Analyses*. Cambridge, MA: National Bureau of Economic Research. https://doi.org/10.3386/w4897.

Brooks, N. 2022. *Environmental Justice Framework*. Minnesota Pollution Control Agency. https://www.pca.state.mn.us/sites/default/files/p-gen5-05.pdf.

Budolfson, M., F. Dennig, F. Errickson, S. Feindt, M. Ferranna, M. Fleurbaey, D. Klenert, et al. 2021. "Climate Action with Revenue Recycling Has Benefits for Poverty, Inequality and Well-Being." *Nature Climate Change* 11(12):1111–1116. https://doi.org/10.1038/s41558-021-01217-0.

Carraro, C., M. Galeotti, and M. Gallo. 1996. "Environmental Taxation and Unemployment: Some Evidence on the "Double Dividend Hypothesis" in Europe." *Journal of Public Economics* 62(1–2):141–181. https://doi.org/10.1016/0047-2727(96)01577-0.

CEQ (Council on Environmental Quality). 2021. *Report to Congress on Carbon Capture, Utilization, and Sequestration*. Washington, DC. https://www.whitehouse.gov/wp-content/uploads/2021/06/CEQ-CCUS-Permitting-Report.pdf.

CEQ. 2022. "Carbon Capture, Utilization, and Sequestration Guidance." *Federal Register* 87(32):8808–8811.

Climate Justice Alliance. 2022. "Environmental Justice Organizations Post Comments on Carbon Capture and Storage to the White House Council on Environmental Quality Indigenous Environmental Network." https://www.ienearth.org/environmental-justice-organizations-post-comments-on-carbon-capture-and-storage-to-the-white-house-council-on-environmental-quality.

Clinton, W.J. 1993. "Executive Order 12866: Regulatory Planning and Review." *Federal Register* 58(190):51735–51744.

Clinton, W.J. 1994. "Executive Order 12898: Federal Actions to Address Environmental Justice in Minority Populations and Low-Income Populations." *Federal Register* 59(32):1–5. https://www.archives.gov/files/federal-register/executive-orders/pdf/12898.pdf.

Cochran, J., and P. Denholm, eds. 2021. *LA100: The Los Angeles 100% Renewable Energy Study*. NREL/TP-6A20-79444. Golden, CO: National Renewable Energy Laboratory. https://maps.nrel.gov/la100.

DOE (U.S. Department of Energy). 2022. "Biden-Harris Administration Launches $2.6 Billion Funding Programs to Slash Carbon Emissions." Energy.gov. https://www.energy.gov/articles/biden-harris-administration-launches-26-billion-funding-programs-slash-carbon-emissions?utm_medium=email&utm_source=govdelivery.

DOE-NE (Office of Nuclear Energy). n.d. "Consent-Based Siting." Energy.gov. https://www.energy.gov/ne/consent-based-siting.

DOE-OEID (Office of Economic Impact and Diversity). n.d. "Justice40 Initiative." Energy.gov. https://www.energy.gov/diversity/justice40-initiative.

EFI (Energy Futures Initiative). 2020. "An Action Plan for Carbon Capture and Storage in California: Opportunities, Challenges, and Solutions." https://energyfuturesinitiative.org/reports/an-action-plan-for-carbon-capture-and-storage-in-california.

Emory University. n.d. "Common Terms. University Office of Diversity, Equity, and Inclusion." https://equityandinclusion.emory.edu/resources/self-guided-learning/common-terms.html.

Energy Star. 2022. "About Energy Star—2021." https://www.energystar.gov/about.

EPA (U.S. Environmental Protection Agency). 2021a. "Learn About Environmental Justice." https://www.epa.gov/environmentaljustice/learn-about-environmental-justice.

EPA. 2021b. "Public Participation Guide." Updated July 12. https://www.epa.gov/international-cooperation/public-participation-guide.

Flores-Jones, I. 2022. "Environmental Justice Groups Call for Full, Coordinated Phase Out of Fossil Fuels by 2045 in California's Climate Change Scoping Plan." California Environmental Justice Alliance. https://caleja.org/2022/03/environmental-justice-groups-call-for-full-coordinated-phase-out-of-fossil-fuels-by-2045-in-californias-climate-change-scoping-plan.

Forbes, S.M., P. Verma, T.E. Curry, S.J. Friedmann, and S.M. Wade. 2008. *Guidelines for Carbon Dioxide Capture, Transport and Storage*. Washington, DC: World Resources Institute.

Goulder, L.H. 1995. "Environmental Taxation and the Double Dividend: A Reader's Guide." *International Tax and Public Finance* 2(2):157–183. https://doi.org/10.1007/BF00877495.

Goulder, L.H., and M.A.C. Hafstead. 2018. *Confronting the Climate Challenge: U.S. Policy Options*. New York: Columbia University Press.

Goulder, L.H., and A.R. Schein. 2013. "Carbon Taxes Versus Cap and Trade: A Critical Review." *Climate Change Economics* 4(3):1350010. https://doi.org/10.1142/S2010007813500103.

Goulder, L., M.A. Hafstead, G. Kim, and X. Long. 2018. *Impacts of a Carbon Tax Across US Household Income Groups: What Are the Equity-Efficiency Trade-Offs?* Cambridge, MA: National Bureau of Economic Research. https://doi.org/10.3386/w25181.

Griliches, Z. 1957. "Hybrid Corn: An Exploration in the Economics of Technological Change." *Econometrica* 25(4):501. https://doi.org/10.2307/1905380.

Hawkins, J., A. Duguid, and L. Keister. 2021. "CO_2 Pipeline Risk Assessment for a Regional-Scale Pipeline in the Midcontinental United States." In *Proceedings of the 15th Greenhouse Gas Control Technologies Conference 15–18 March 2021*. https://dx.doi.org/10.2139/ssrn.3821323.

IEP (Internet Encyclopedia of Philosophy). n.d. "The American Environmental Justice Movement." https://iep.utm.edu/enviro-j.

IIJA (Infrastructure Investment and Jobs Act). Congress. 2021. Public Law 117-58, § 40305. H.R.3684 - Infrastructure Investment and Jobs Act. Public Law 117-58. 117th Congress (2021–2022). https://www.congress.gov/bill/117th-congress/house-bill/3684/text.

Initiative for Energy Justice. 2019. *The Energy Justice Workbook*. Boston, MA. https://iejusa.org/wp-content/uploads/2019/12/The-Energy-Justice-Workbook-2019-web.pdf.

IWG (Interagency Working Group on Social Cost of Greenhouse Gases, U.S. Government. 2021. *Technical Support Document: Social Cost of Carbon, Methane, and Nitrous Oxide Interim Estimates under Executive Order 13990*. Washington, DC. https://www.whitehouse.gov/wp-content/uploads/2021/02/TechnicalSupportDocument_SocialCostofCarbonMethaneNitrousOxide.pdf.

Justice40. 2021. "Homepage." https://www.thejustice40.com.

Kavlak, G., J. McNerney, and J.E. Trancik. 2018. "Evaluating the Causes of Cost Reduction in Photovoltaic Modules." *Energy Policy* 123(December):700–710. https://doi.org/10.1016/j.enpol.2018.08.015.

Kosar, U., and V. Suarez. 2021. "Removing Forward: Centering Equity and Justice in a Carbon-Removing Future." *Carbon180* August.

LEP (Labor Energy Partnership). 2021. *Building to Net-Zero: A U.S. Policy Blueprint for Gigaton-Scale CO_2 Transport and Storage Infrastructure*. Washington, DC: Energy Futures Initiative and AFL-CIO. https://laborenergy.org/wp-content/uploads/2021/10/LEP-Building_to_Net-Zero-June-2021-v4.pdf.

Mansfield, E. 1996. *Estimating Social and Private Returns from Innovations Based on the Advanced Technology Program: Problems and Opportunities*. Technical Report. Gaithersburg, MD: National Institute of Standards and Technology. https://nvlpubs.nist.gov/nistpubs/gcr/1999/gcr99-780.pdf.

McDonald, A., and L. Schrattenholzer. 2001. "Learning Rates for Energy Technologies." *Energy Policy* 29(4):255–261. https://doi.org/10.1016/S0301-4215(00)00122-1.

Metcalf, G.E. 2019. "On the Economics of a Carbon Tax for the United States." *Brookings Papers on Economic Activity* (1):405–484. https://doi.org/10.1353/eca.2019.0005.

NGA (Natural Gas Act). 2006. 15 U.S.C. 717.

Nielsen, J.A.E., K. Stavrianakis, and Z. Morrison. 2022. "Community Acceptance and Social Impacts of Carbon Capture, Utilization and Storage Projects: A Systematic Meta-Narrative Literature Review." *PLOS ONE* 17(8):e0272409. https://doi.org/10.1371/journal.pone.0272409.

NJDEP (New Jersey Department of Environmental Protection). 2022. "Environmental Justice Rules." *New Jersey Register* 54(11). https://www.nj.gov/dep/rules/proposals/proposal-20220606a.pdf.

Nordhaus, W. 2011. Designing a Friendly Space for Technological Change to Slow Global Warming." *Energy Economics* 33(4):665–673. https://doi.org/10.1016/j.eneco.2010.08.005.

Nordhaus, W. 2019. "Climate Change: The Ultimate Challenge for Economics." *American Economic Review* 109(6):1991–2014. https://doi.org/10.1257/aer.109.6.1991.

Parry, I.W.H. 1995. "Pollution Taxes and Revenue Recycling." *Journal of Environmental Economics and Management* 29(3):S64–S77. https://doi.org/10.1006/jeem.1995.1061.

Parry, I.W.H. 1998. "A Second-Best Analysis of Environmental Subsidies." *International Tax and Public Finance* 5(2):153–170. https://doi.org/10.1023/A:1008638320593.

Parry, I., A. Morris, and R.C. Williams III, eds. 2015. *Implementing a US Carbon Tax*. New York: Routledge. https://doi.org/10.4324/9781315747682.

Parry, I., S. Black, and K. Zhunussova. 2022. *Carbon Taxes or Emissions Trading Systems? Instrument Choice and Design*. IMF Staff Climate Note 2022/006. Washington, DC: International Monetary Fund.

PHMSA (Pipeline and Hazardous Materials Safety Administration). 2019. "Pipeline Enforcement Guidance." U.S. Department of Transportation. https://www.phmsa.dot.gov/pipeline/enforcement/enforcement-program-0.

PHMSA. 2022. "PHMSA Announces New Safety Measures to Protect Americans from Carbon Dioxide Pipeline Failures After Satartia, MS Leak." U.S. Department of Transportation. https://www.phmsa.dot.gov/news/phmsa-announces-new-safety-measures-protect-americans-carbon-dioxide-pipeline-failures.

Popp, D. 2002. "Induced Innovation and Energy Prices." *American Economic Review* 92(1):160–180. https://doi.org/10.1257/000282802760015658.

Principles of Environmental Justice. 1996. "Environmental Justice/Environmental Racism." http://www.ejnet.org/ej/principles.html.

Romer, P.M. 1990. "Endogenous Technological Change." *Journal of Political Economy* 98(5, Part 2):S71–S102. https://doi.org/10.1086/261725.

Romero-Lankao, P. 2022. "Environmental Justice Considerations for CO_2 Utilization." Presentation to the National Academies' Committee on Carbon Utilization Infrastructure, Markets, Research and Development. March 3. https://www.nationalacademies.org/event/03-01-2022/carbon-utilization-infrastructure-markets-research-and-development-meeting-2.

Samuelson, P.A. 1954. "The Pure Theory of Public Expenditure." *Review of Economics and Statistics* 36(4):387. https://doi.org/10.2307/1925895.

Stavins, R.N. 2019. *Carbon Taxes vs Cap and Trade: Theory and Practice*. Discussion Paper ES 2019-9. Cambridge, MA: Harvard Project on Climate Agreements.

Stavins, R.N. 2020. "The Future of US Carbon-Pricing Policy." *Environmental and Energy Policy and the Economy* 1(January):8–64. https://doi.org/10.1086/706792.

Stoneman, P., and G. Battisti. 2010. "Chapter 17 - The Diffusion of New Technology." Pp. 733–760 in Vol. 2 of *Handbook of the Economics of Innovation*, B.H. Hall and N. Rosenberg, eds. Amsterdam, Netherlands: North Holland. https://doi.org/10.1016/S0169-7218(10)02001-0.

Sunstein, C.R. 2014. "The Real World of Cost-Benefit Analysis: Thirty-Six Questions (and Almost as Many Answers)." *Columbia Law Review* 114(1):167–211.

Sunstein, C.R. 2017. "Cost-Benefit Analysis and Arbitrariness Review." *SSRN Electronic Journal*. https://doi.org/10.2139/ssrn.2752068.

Tietenberg, T.H., and L. Lewis. 2018. *Environmental and Natural Resource Economics*, 11th ed. New York: Routledge.

Tollefson, L., S. Greenberg, J. Bradbury, L. Cumming, D. Daly, G. Garrett, M. Cather, R. Myhre, M. Stone, and S. Wade. 2017. *Best Practices: Public Outreach and Education for Geologic Storage Projects*. Pittsburgh, PA: National Energy Technology Laboratory. https://netl.doe.gov/sites/default/files/2018-10/BPM_PublicOutreach.pdf.

TPU (Tacoma Public Utilities). 2021. "Tacoma Power Announces the Nation's First Electrofuel Tariff." https://www.mytpu.org/tacoma-power-announces-the-nations-first-electrofuel-tariff.

Verdolini, E., and M. Galeotti. 2011. "At Home and Abroad: An Empirical Analysis of Innovation and Diffusion in Energy Technologies." *Journal of Environmental Economics and Management* 61(2):119–134. https://doi.org/10.1016/j.jeem.2010.08.004.

6

Priority Infrastructure Opportunities for CO_2 Utilization

Building on the analyses of carbon dioxide (CO_2)-derived products, infrastructure requirements, and policy, regulatory, and societal considerations discussed in Chapters 2 through 5, this chapter presents a summary of priority infrastructure opportunities to enable CO_2 utilization. The chapter begins by describing options for CO_2 utilization infrastructure funding based on current policy and regulatory regimes, and considering successful examples in related industries. It then examines near-term opportunities for CO_2 utilization infrastructure investments, as well as near-term actions to enable longer-term deployment options. A primary consideration for these opportunities is the ability of CO_2 utilization to participate in a future circular carbon economy, which depends on the type of CO_2 source, CO_2-derived product lifetime, and life cycle emissions of other process inputs. The chapter ends with findings and recommendations focused on implementing these identified opportunities.

6.1 INFRASTRUCTURE FUNDING AND INVESTMENTS

When thinking about infrastructure funding, the first consideration is to understand what is allowed or incentivized by policy or law, and then to understand the economic implications of different options. Minimization of cost per unit mass CO_2 mitigated is an important metric of success to motivate CO_2 capture and utilization development broadly and marketable products specifically. Credible, product-level demonstration of cost minimization using captured CO_2 has proven the basis for demand for such a product in a carbon-constrained world. Provided with this framing, the next considerations at a most basic level are the total unit cost to fabricate a product utilizing captured CO_2 and obtaining sufficient revenue to cover such costs (and a fair return).

The most resource-efficient approach to building infrastructure for CO_2 utilization is to consider the entire value chain at the same time, given the need to align many of the individual links along the chain. A successful example of this approach is the Alberta Carbon Trunk Line (ACTL) located in Canada, which mapped out CO_2 supply, transportation, and disposition as part of an integrated project (Alberta Government 2020). The ACTL links together capture from industrial sources in one part of the province, conveyed along an (initially oversized) pipeline to geologic sequestration and enhanced oil recovery sites in another. One impetus for the project, among others, was the downstream demand for the CO_2 and the revenue that it would generate. This framing is useful when considering the context for CO_2 utilization.

To assess the entire value chain of carbon utilization projects, one would begin by estimating revenue from demand for the products created by CO_2 conversion and the expected selling price for such products. Total input costs include the cost of capture and transportation of CO_2 and any other costs associated with enabling infrastructure. Project finance will be enabled when revenues are sufficiently large, consistent, and of long enough duration to pay down all capital and provide a fair return on top of the cost of ongoing operations. Incentives and contracting tools can serve as remedies where revenues fall short of these dimensions, either due to initial weak market demand or revenues insufficient to make the project of rated investment quality. For example, the 45Q or 45V tax credits would lower the input costs of CO_2 and hydrogen, respectively, thereby increasing the possibility of revenue sufficiency. Moreover, a procurement mechanism using a take-or-pay contract mechanism set at a sufficient price would guarantee the revenue irrespective of market fluctuations, thereby lowering the cost of capital associated with the project, since the probability of capital repayment is enhanced. Liquefied natural gas terminal project development uses this type of contracting mechanism. In either case—reducing input costs through tax incentives or other support mechanisms or enhancing revenue through procurement mechanisms or contracting terms that reduce volatility and/or offset required sales premiums—while public policy may provide the incentives, it is fully within the realm of the private sector to build, own, and operate the conversion facility. The same is true for a capture facility, provided there is sufficient infrastructure to move the CO_2 from capture to the point of utilization.

As articulated in Section 4.3, the most cost-effective transportation option is via pipeline, depending on the distance and volume of CO_2 to be moved. Setting aside the need to obtain right-of-way, easement, and other relevant permits, developing a pipeline is predicated on securing adequate commodity flow via throughput agreements. However, there is a trade-off between constructing a pipeline that can handle immediate volumetric needs versus a larger pipeline that is otherwise overbuilt, given current market volumes compared to potential future demands. For cases that can credibly demonstrate that future demand will exceed current transportation needs, public-sector support of pipeline buildout may be advantageous. Public funding would remove the volume and timing risk associated with uncertain market developments and serve as an important catalyst for driving supply of and demand for captured CO_2. Although within the scope of the electricity industry, the Texas Competitive Renewable Energy Zone (CREZ) is a good example of public-sector vision, planning, and development of infrastructure at a scale meant to support an expanding market (Cohn and Jankovska 2020). In this case, the Public Utility Commission of Texas planned and brokered the financing for the development of CREZ, where transmission developers earned return through payments from electricity ratepayers. In the case of CO_2 pipelines, similar to what occurred in the ACTL example, the public entity can finance the project development and retain the right to sell off the asset to a private entity once a specified portion of the volumetric capacity has been attained. As discussed in Section 5.2.4, Section 40304 of the Infrastructure Investment and Jobs Act (IIJA) authorizes grants and loans for building CO_2 common carrier infrastructure through the Carbon Dioxide Transportation Infrastructure Finance and Innovation Program. The loans aim to help eligible projects attract investment and begin earlier than would otherwise be possible, while the grants target the costs of constructing a facility that can accommodate future growth in demand for CO_2 transport. The IIJA calls for the Secretary of Energy to prioritize projects that are large-capacity, common-carrier infrastructure, have demonstrated demand for infrastructure from CO_2 capture facilities, represent geographic diversity, and site infrastructure within existing corridors to minimize environmental disturbance and other siting concerns (IIJA 2021, § 40305).

Of course, where practical, the capture and production facilities could be both "behind the fence," such that any kind of CO_2 transportation could proceed over short distances and minimize disturbances to surrounding lands. In this case, need for any direct public investment is unlikely; however, this kind of point-to-point pipeline may be limited in volume and perhaps may be less cost-efficient in the event that there needs to be an expansion. An example of this kind of layout may be that of a fertilizer production facility, which operates a steam methane reforming process to make hydrogen and captures some of the emitted CO_2 for use in the production of urea. Two obvious opportunities to deploy CO_2 utilization facilities are within the hydrogen and direct air capture (DAC) hubs designated to receive significant funding via appropriations associated with the IIJA (§§ 40308, 40314). Another opportunity stemming from the IIJA is the U.S. Department of Energy's (DOE's) Carbon Dioxide Transport/Front-End Engineering Design Program, which aims to "design regional carbon dioxide pipeline systems to safely transport CO_2 from key sources to centralized locations" and for which DOE issued a notice of intent for funding in July 2022 (DOE 2022).

Finally, in fashion similar to pipeline transportation, geologic sequestration sites benefit from economies of scale, where it would be cost-efficient to construct a large repository (and necessary monitoring, reporting, and verification infrastructure) that could receive injections from multiple sources. Therefore, it would be beneficial for a public entity to lead in sponsoring site selection, characterization, and construction, in anticipation of greater volumes of CO_2 needing sequestration ultimately materializing. IIJA § 40305 includes authorization and funding that expands an existing DOE program for carbon storage validation and testing to include commercialization and associated CO_2 transport infrastructure. The commercialization program intends to facilitate new or expanded carbon capture and storage (CCS) and associated CO_2 transport infrastructure, including funding for feasibility, site characterization, permitting, and construction, giving priority to those storing substantial amounts of CO_2, or those collecting from multiple capture facilities (IIJA 2021, § 40305). Depending on demand projections, the revenue model that supports the project finance could be based on either a capacity payment or a tolling/tipping fee payment arrangement. The latter of these two approaches is similar to the municipal waste model; however, the waste model is based on relatively stable usage rates (i.e., number of trucks entering and tipping waste to the landfill per unit of time).

6.2 NEAR-TERM VERSUS LONG-TERM INFRASTRUCTURE STRATEGIES

Expanding infrastructure to meet climate goals and carbon management objectives will be a major challenge requiring deliberate, long-term planning. Given the smaller potential scale of CO_2 utilization relative to other CO_2 emissions reduction and carbon management opportunities, it will not be a driving factor in large infrastructure development, but rather additive to ongoing infrastructure projects that capture CO_2. In the near term, opportunities for investable projects in CO_2 utilization that align with a future circular carbon economy are limited (see, e.g., Bazzanella and Ausfelder 2017; Centi et al. 2020; Gabrielli et al. 2020; Hepburn et al. 2019; Soler 2020). Sources of CO_2 are likely to change significantly over the next 30 years, with many point sources of CO_2, such as fossil fueled power plants, being phased out, and new sources, such as DAC, developing. This will complicate investments that require 20+ year facility lifetimes to yield returns. Currently, the cost of manufacturing hydrocarbon products from CO_2, hydrogen, and electricity exceeds that of existing manufacturing processes significantly due to the high hydrogen consumption required and capital intensity of conversion steps (Frontier Economics 2018; Huang et al. 2021). Costs can be even higher if net-zero emissions CO_2 utilization is attempted such as through use of clean hydrogen, clean electricity, and CO_2 from DAC, and premiums for "green" products or CO_2 abatement costs are not universal. For these reasons, opportunities for CO_2 utilization need to be considered on a project-by-project basis, and addition of any necessary infrastructure considered as part of the project cost.

Two product pathways are prime targets for early CO_2 utilization infrastructure investments. First is use of biogenic CO_2 for hydrocarbon production, particularly sustainable aviation fuels. CO_2 sourced from bioethanol plants is highly concentrated and biogenic, with a low cost of capture. The scale is small per bioethanol plant, but, when aggregated via pipelines, could produce a sustainable CO_2 source at a scale that can allow production of synthetic aviation fuel or sustainable chemicals when combined with clean hydrogen. Synthetic aviation fuel, a prime product target, can be transported to market via truck or rail, given the scale of manufacture, and may not require a dedicated liquid pipeline. If synthetic natural gas is the CO_2 utilization product, then existing natural gas pipelines can be used for product transportation. The current market for bioethanol production is blending with gasoline for liquid transportation fuels. In a decarbonized future, this market my decrease substantially; however, bioethanol's use may pivot to other needs for sustainable substitutes for fossil carbon, such as for heavy-duty transportation, as a seasonal store of energy, or as a feedstock for ethylene used in chemical synthesis, potentially maintaining these sources of high-quality CO_2.

A second prime CO_2 utilization opportunity is generating mineral products for construction. Fossil CO_2 point sources can be used in this case, since the mineral products entail long-term sequestration, often into building products. The other required feedstock is mined minerals, which can be high cost if transported over long distances, such that co- or near-location of mining operations with CO_2 capture options is a key consideration. Otherwise, CO_2 pipelines would be required to couple mineral and CO_2 feeds, or mineral transport will be needed (trucks or rail). Similarly, the product (e.g., aggregate for concrete) is a solid and requires transport by truck or rail to the end-use location, so location near demand centers such as urban areas is preferred.

For both of these near-term use cases, infrastructure needs and siting have to be considered on a project or hub basis with input and buy-in from local communities. There is no one-size-fits-all infrastructure model. Nonetheless, locating enabling infrastructure such as clean hydrogen and clean electricity in the proximity of both fossil-derived (i.e., unsustainable) and DAC, direct ocean capture (DOC) and/or biogenic (i.e., sustainable) CO_2 sources could prove cost-effective, as it would enable the manufacturing of sustainable products in both the near and long term.

In addition to the near-term opportunities described above, additional steps may be taken to position CO_2 utilization for future viability. For example, focusing in the near term on decarbonizing the grid and scaling up clean hydrogen production will benefit CO_2 utilization in the long term. Concurrently, continued research and development to reduce the costs and improve the energy and material efficiency of CO_2 capture and conversion technologies, as well as to demonstrate scale-up and establish the commercial viability of new products developed, is needed to position CO_2 utilization for successful deployment as a sustainable option for mitigating climate change. Furthermore, although initial markets for CO_2-derived products may be small, they provide an opportunity for learning on regulatory and institutional issues related to the development and deployment of new products and processes. Net CO_2 emissions across the energy system can be reduced via combinations of learning by doing, reduced carbon intensity of energy and supply chains, and public and commercial acceptance of new technologies and products. The emerging potential of CO_2 utilization needs to be considered in siting of pipelines and infrastructure for carbon capture and storage, renewable energy, and hydrogen production. Also important to consider is the overlap of existing industrial chemicals and materials manufacturing facilities and workforce skills, which in many cases could be repurposed for CO_2 utilization deployments. Identification and support of commercial hubs to enable CO_2 utilization experimentation, development, and de-risking will be critical for successful leveraging of CO_2 utilization opportunities.

In the longer term, DAC or biogenic CO_2, biomass carbon, and chemical circularity (recycling), are anticipated to be the primary sources of carbon for manufacturing. Of these, only biomass carbon and biogenic CO_2 sources (i.e., bioethanol plants) exist today at commercial scale. Planning for future deployment of CO_2 utilization processes requires considering a variety of infrastructure options and requirements. For example, siting for DAC is more likely to need to incorporate infrastructure to transport the CO_2-derived product rather than the CO_2. Optimization of enabling infrastructure is a function of the scale of the manufacturing plant, and there are requirements for constant operation and capacity factor in meeting target economics. Given the large amounts of electricity and/or hydrogen required to upgrade CO_2 to hydrocarbon products, co-locating hydrogen generation with the manufacturing plant utilizing CO_2 as feedstock may be desirable. These decisions will have to be made strategically as integrated hubs are designed, and by following best practices for community engagement as discussed in Chapter 5. In doing so, it must be recognized that the use of CO_2 as a feedstock for future synthetic fuels or chemicals is not a given; bio-based feedstocks and/or chemical recycling may compete with CO_2 capture and utilization pathways.

Specific end uses and projects for CO_2 utilization may be unsustainable relative to better use of renewable or clean electricity, or alternative means for providing energy services to society. CO_2 captured from point sources requires more energy and/or hydrogen to reform to fuels than replacement of those fuels by direct electrification (via clean energy) or hydrogen, where the latter are "zero emission"[1] and do not contribute to local air pollution and equity issues in human health (de Kleijne et al. 2022; Mac Dowell et al. 2017; Serdoner and Whiriskey 2017; Soler 2020; Yugo and Soler 2019). Reuse of fossil CO_2 to make fuels or other short-lived chemical products, which re-release CO_2 upon combustion, degradation, or incineration,[2] at best results in a 50 percent reduction in CO_2 emissions per unit of energy service for the use of the fuel originally and again with one-time recycling. In reality, the net emissions reduction is much less (20 percent) due to conversion inefficiencies.[3] The greenhouse gas (GHG)

---

[1] Hydrogen combustion can lead to criteria pollutant thermal NO_x if not done in a judicious manner.

[2] Hydrocarbon products derived from CO_2 have an average effective sequestration of less than 100 years despite the fact that they do not chemically degrade for hundreds or even thousands of years. This is because product lives are shorter than 100 years, and a substantial portion of end-of-life waste is incinerated—20–80 percent today, and likely a higher percentage in the future given space constraints, environmental runoff issues, and a desire for energy recovery.

[3] The combustion of a hydrocarbon fuel yields the lower heating value of energy release and the associated CO_2 emissions. If renewable energy is used to capture the emitted CO_2 and convert it back into a hydrocarbon fuel, and then the fuel is combusted again, the same net emissions result but with twice the energy output. However, that 50 percent reduction in CO_2 footprint assumes that capture and conversion are 100 percent efficient, so in reality, the CO_2 footprint will be reduced by less than 50 percent compared to the "no recycling" case.

reduction benefit can be improved by using biogenic CO_2 from bioethanol fermentation or CO_2 from DAC or DOC as the carbon feedstock for synthesizing fuels or short-lived chemicals. However, particularly for the case of fuels, if used for local transport in urban areas, they still create the same air pollution inequities that arise from use of fossil fuels today. Synthetic aviation fuel is one sustainable and equitable option for biogenic CO_2 use, given that aviation's primary deployment is distant from population. Chemical and fuel production pathways must compete with the use of bio-based feedstocks (e.g., biomass) to make chemicals for a future circular chemical economy, where the added energy and unit costs may be lower (Lange 2021). For these reasons, a full life cycle assessment of any proposed CO_2 utilization project is critical to ensure effective use of capital and renewable or clean energy resources for addressing climate change and avoiding propagation of air pollution stresses on communities.

6.3 FINDINGS AND RECOMMENDATIONS ON PRIORITY INFRASTRUCTURE OPPORTUNITIES FOR CO_2 UTILIZATION

FINDING 6.1 Near-Term Opportunities for CO_2 Utilization. Options for near-term deployment of CO_2 utilization exist and can be identified via a combination of techno-economic and life cycle assessments. One example is low-cost capture of highly concentrated biogenic CO_2 fermentation exhaust gas from ethanol production, which, when combined with low- or zero-carbon-emission hydrogen, can produce sustainable chemicals or fuels for heavy-duty transportation such as shipping and aviation. Another example is the exothermic reaction of minerals with CO_2 to form mineral carbonates for the building/construction industry. However, it is also possible to utilize CO_2 in costly and unsustainable ways that will result in increased fossil CO_2 emissions, environmental damage, and societal injustice relative to competing options, if the appropriate systems-level analyses are not considered.

RECOMMENDATION 6.1. The U.S. Department of Energy should support its national laboratories, academia, and industry to leverage their competencies in techno-economic and life cycle assessments, as well as integrated systems analysis, to identify the best deployment and investment opportunities from the myriad of utilization options, avoiding those that are technically feasible but not sustainable or economically attractive. These assessments should consider relevant regulatory and policy frameworks and environmental justice impacts, as well as factors that may influence societal acceptance of the technologies.

FINDING 6.2 Flexible Infrastructure for CO_2 Capture, Utilization, and Storage. Carbon dioxide capture and transportation infrastructure for sequestration also may serve utilization projects, depending on the type, purity, and location of the CO_2 source; the utilization product; and the other energy and feedstock requirements. For example, fossil CO_2 sources are only sustainable in a net-zero future for utilization into durable products, for example, concrete, aggregates, and carbon fiber, unless utilization into a short-lived product is paired with a separate, verifiable negative-emissions process (e.g., direct air capture plus storage). In general, the ability to divert some of the CO_2 destined for storage to a utilization process provides an economic driver for infrastructure development, exploiting economies of scale where possible and economies of numbers where needed, and may in some cases improve public perception of carbon management practices.

RECOMMENDATION 6.2. The U.S. Department of Energy (DOE) should favorably consider the ability to connect to future CO_2 utilization processes and technologies when reviewing CO_2 capture, transport, and storage demonstration projects. Designing such flexibility into the initial infrastructure development could provide long-term benefits, since CO_2 utilization likely will be required to produce carbon-based products in a net-zero emissions future. DOE should document and share learning from such demonstrations with CO_2 utilization and carbon capture and storage project developers to facilitate future CO_2 utilization opportunities.

FINDING 6.3 Industrial Clusters for CO_2 Capture, Utilization, and Storage. Industrial clusters that co-locate capture, utilization, and storage of CO_2 provide the capability for managing large volumes of CO_2 without the need for extensive pipeline networks and have the flexibility to incorporate different utilization processes if market trends change or new technologies are developed over time. Additionally, locating such carbon capture, utilization, and storage clusters in regions that already have a large industrial presence would maintain jobs, knowledge, and workforce, allowing know-how to be recycled into a new, related industry. However, the CO_2 sources in *current* industrial clusters are primarily fossil-carbon based, which is not desirable or sustainable for all utilization applications.

> **RECOMMENDATION 6.3.** As part of its industrial decarbonization efforts, the U.S. Department of Energy should provide technical and financial support for the development of industrial clusters for carbon capture, utilization, and storage (CCUS) that capture CO_2 in large amounts and include the necessary infrastructure for both utilization and storage of CO_2. CCUS cluster development should involve best practices for community engagement and allow for flexibility in utilization scenarios over the long term, for example, by incorporating hydrogen production, chemical and fuel manufacturing, and low-carbon electricity generation. To achieve sustainability goals, these clusters should route the majority of CO_2 captured from fossil sources to long-term geologic storage or production of durable CO_2 products (e.g., mineralization products, carbon fiber, and other solid carbon materials). Infrastructure for producing nondurable CO_2-derived products (e.g., chemicals and fuels) should incorporate CO_2 from direct air capture or biogenic sources where possible, or be paired with verifiable negative-emissions projects to offset fossil CO_2 use.

> **RECOMMENDATION 6.4.** When evaluating proposals for the hydrogen and direct air capture hubs authorized in the Infrastructure Investment and Jobs Act, the U.S. Department of Energy should consider rewarding through their selection process projects that co-locate hub types to take advantage of shared infrastructure needs and facilitate CO_2 utilization applications that require hydrogen.

FINDING 6.4 Near-Term Opportunities for Maximum Climate Benefit via Strategic Co-location to Minimize Transport. Strategic co-location that considers the features of CO_2 sources and utilization products can maximize climate benefits and minimize transportation requirements for near-term CO_2 utilization opportunities. In particular, CO_2 utilization facilities making durable, solid carbon products co-located with fossil, biogenic, or direct air capture CO_2 sources near urban demand centers could enable net-zero or net-negative manufacturing of building materials like cement and aggregates while minimizing transport of heavy, high-volume products. Likewise, co-location of direct air capture or biogenic CO_2 sources with utilization facilities making liquid products like sustainable aviation fuels near existing infrastructure for enabling inputs and product transport could produce net-zero emissions fuels while minimizing transport of the CO_2 feedstock.

> **RECOMMENDATION 6.5.** The U.S. Department of Energy should work with its national laboratories, university researchers, and industry partners to conduct detailed studies that identify the most promising opportunities for CO_2 utilization infrastructure based on technological, environmental, economic, and societal factors. These studies should examine opportunities to (1) co-locate sources of CO_2, utilization facilities, and product users to minimize transport and (2) site utilization facilities in proximity to existing transport and delivery infrastructure. The studies should determine the value of co-locating specific CO_2 utilization activities with specific source types of CO_2, as well as the value of minimizing transport, identifying those that maximize climate benefits, either net-negative or net-zero.

6.4 REFERENCES

Alberta Government. 2020. "*Alberta Carbon Trunk Line Project: Knowledge Sharing Report, 2019*. Alberta, CA. https://open.alberta.ca/publications/alberta-carbon-trunk-line-project-knowledge-sharing-report-2019.

Bazzanella, A.M., and F. Ausfelder. 2017. *Low Carbon Energy and Feedstock for the European Chemical Industry*. Technology Study. Germany: DECHEMA. https://dechema.de/dechema_media/Downloads/Positionspapiere/Technology_study_Low_carbon_energy_and_feedstock_for_the_European_chemical_industry-p-20002750.pdf.

Centi, G., S. Perathoner, A. Salladini, and G. Iaquaniello. 2020. "Economics of CO_2 Utilization: A Critical Analysis." *Frontiers in Energy Research* 8. https://www.frontiersin.org/article/10.3389/fenrg.2020.567986.

Cohn, J., and O. Jankovska. 2020. *Texas CREZ Lines: How Stakeholders Shape Major Energy Infrastructure Projects*. Houston, TX: Rice University Baker Institute for Public Policy. https://www.bakerinstitute.org/research/texas-crez-lines-how-stakeholders-shape-major-energy-infrastructure-projects.

de Kleijne, K., S.V. Hanssen, L. van Dinteren, M.A.J. Huijbregts, R. van Zelm, and H. de Coninck. 2022. "Limits to Paris Compatibility of CO_2 Capture and Utilization." *One Earth* 5(2):168–185. https://doi.org/10.1016/j.oneear.2022.01.006.

DOE (U.S. Department of Energy). 2022. "Biden-Harris Administration Launches $2.6 Billion Funding Programs to Slash Carbon Emissions." Energy.gov. https://www.energy.gov/articles/biden-harris-administration-launches-26-billion-funding-programs-slash-carbon-emissions?utm_medium=email&utm_source=govdelivery.

Frontier Economics. 2018. *The Future Cost of Electricity-Based Synthetic Fuels*. Berlin: Agora Energiewende. https://static.agora-energiewende.de/fileadmin/Projekte/2017/SynKost_2050/Agora_SynKost_Study_EN_WEB.pdf.

Gabrielli, P., M. Gazzani, and M. Mazzotti. 2020. "The Role of Carbon Capture and Utilization, Carbon Capture and Storage, and Biomass to Enable a Net-Zero-CO_2 Emissions Chemical Industry." *Industrial & Engineering Chemistry Research* 59(15):7033–7045. https://doi.org/10.1021/acs.iecr.9b06579.

Hepburn, C., E. Adlen, J. Beddington, E.A. Carter, S. Fuss, N. Mac Dowell, J.C. Minx, P. Smith, and C. K. Williams. 2019. "The Technological and Economic Prospects for CO_2 Utilization and Removal." *Nature* 575(7781):87–97. https://doi.org/10.1038/s41586-019-1681-6.

Huang, Z., R.G. Grim, J.A. Schaidle, and L. Tao. 2021. "The Economic Outlook for Converting CO_2 and Electrons to Molecules." *Energy & Environmental Science* 14(7):3664–3678. https://doi.org/10.1039/D0EE03525D.

IIJA (Infrastructure Investment and Jobs Act). 2021. "H.R.3684 - Infrastructure Investment and Jobs Act." Public Law 117-58, § 40305. 117th Congress (2021–2022). https://www.congress.gov/bill/117th-congress/house-bill/3684/text.

Lange, J.-P. 2021. "Towards Circular Carbo-Chemicals—the Metamorphosis of Petrochemicals." *Energy & Environmental Science* 14(8):4358–4376. https://doi.org/10.1039/D1EE00532D.

Mac Dowell, N., P.S. Fennell, N. Shah, and G.C. Maitland. 2017. "The Role of CO_2 Capture and Utilization in Mitigating Climate Change." *Nature Climate Change* 7(4):243–249. https://doi.org/10.1038/nclimate3231.

Serdoner, A., and K. Whiriskey. 2017. *The "Power to Liquids" Trap*. Brussels: Bellona Europa. https://network.bellona.org/content/uploads/sites/3/2017/04/Power-to-Liquids_BellonaEuropa-1.pdf.

Soler, A. 2020. *Role of E-Fuels in the European Transport System—Literature Review*. Report No. 14/19. Brussels: Concawe. https://www.concawe.eu/wp-content/uploads/Rpt_19-14.pdf.

Yugo, M., and A. Soler. 2019. "A Look into the Role of E-Fuels in the Transport System in Europe (2030–2050)." *Concawe Review* 28(1):4–22. https://www.concawe.eu/wp-content/uploads/E-fuels-article.pdf.

Appendixes

A

Committee Member Biographies

EMILY A. CARTER (*Chair*) is the Gerhard R. Andlinger Professor in Energy and the Environment and a professor of mechanical and aerospace engineering and applied and computational mathematics at Princeton University. She is also a senior strategic advisor for sustainability science at the U.S. Department of Energy's (DOE's) Princeton Plasma Physics Laboratory. Until the end of 2021, Carter served as the executive vice chancellor and provost (EVCP) and distinguished professor of chemical and biomolecular engineering at the University of California, Los Angeles (UCLA). Carter earned a B.S. in chemistry from the University of California, Berkeley (graduating Phi Beta Kappa), and a Ph.D. in chemistry from the California Institute of Technology, followed by a brief postdoctoral fellowship at the University of Colorado Boulder, before spending 16 years on UCLA's chemistry and biochemistry faculty. She moved to Princeton University in 2004, where she spent 15 years as jointly appointed faculty in mechanical and aerospace engineering and applied and computational mathematics. From 2010 to 2016, she was Princeton's founding director of the Andlinger Center for Energy and the Environment, and from 2016 to 2019, she was Princeton's dean of engineering and applied science before returning to co-lead UCLA as its EVCP. Carter develops and applies quantum mechanical simulation techniques to enable the discovery and design of molecules and materials for sustainable energy, fuels, and chemicals, supported by grants from the U.S. Department of Defense and DOE. She has received numerous honors, including election to the National Academy of Sciences, the American Academy of Arts & Sciences, the National Academy of Inventors, and the National Academy of Engineering.

SHOTA ATSUMI is a professor in the Department of Chemistry at the University of California, Davis. He was a postdoctoral researcher with Dr. John W. Little at the University of Arizona and with Dr. James C. Liao at the University of California, Los Angeles. His current research focuses on the use of synthetic biology and metabolic engineering approaches to engineer microorganisms to convert CO_2 to valuable chemicals. The primary research goals of his group are to develop a platform for valuable chemical production from carbon dioxide using photosynthetic microorganisms and to develop novel biosynthetic pathways to produce chemical compounds that microbes naturally produce in trace amounts or not at all. He received the Hellman Fellowship in 2021, a National Science Foundation CAREER award in 2014, and the Chancellor's Fellowship in 2018. Atsumi received his Ph.D. in biological chemistry in 2002, his M.S. in biological chemistry in 1998, and his B.S. in 1996, all from Kyoto University.

MAKINI BYRON is the director of External Technologies at Linde, a leading industrial gas and engineering company. In her current role, Byron identifies technologies at startups, universities, and research institutes that can benefit her company and gains access through partnerships, collaborations, or investments. She has over 12 years of experience within the energy and chemicals sectors, with a focus on emerging technologies for decarbonization. Byron has managed or participated in several U.S. Department of Energy–funded projects for the commercial engineering design and demonstration of post-combustion and oxy-combustion carbon capture technologies. Her technology development experience includes several CO_2 utilization technologies, from biological conversion of CO_2 to valuable end products, mineralization of CO_2 to cementitious material, and the application of supercritical CO_2 for lubrication and cooling. She has an M.S. in chemical engineering and a certificate in science, technology, and energy policy from Princeton University. Byron is a registered project management professional and a member of the Project Management Institute and the American Institute of Chemical Engineers.

STEPHEN COMELLO is the senior vice president of Strategic Initiatives at the Energy Futures Initiative Foundation and the deputy director of its Energy Futures Finance Forum. He is also a senior research fellow (nonresident) at the Steyer-Taylor Center for Energy Policy and Finance at Stanford University. Previously, he spent close to a decade as the research director of the Energy Business Innovations focus area and as a lecturer in management at the Stanford Graduate School of Business. Through the lenses of techno-economics, public policy, project finance, and business model innovation, his work examines the scale-up opportunities for low-carbon energy and environmental solutions. Comello has been a lead author on various academic publications and government reports exploring topics such as carbon capture and utilization in the industrial sector; at-scale deployment of carbon capture, use, and storage in the United States; policy for carbon capture in California; the market and policy prospects for clean hydrogen industrial hubs in the United States; financial regulation and clean energy capital flows; and corporate decarbonization strategies. He holds a Ph.D. in civil and environmental engineering from Stanford University, and an M.S. and a B.S. in mechanical and industrial engineering from the University of Toronto.

MAOHONG FAN is a School of Energy Resources professor in chemical and petroleum engineering at the University of Wyoming and an adjunct professor in environmental engineering at the Georgia Institute of Technology. He has led and worked on many projects in chemical production, clean energy generation, and environmental protection. The projects have been supported by various domestic and international funding agencies such as the National Science Foundation (NSF), the U.S. Department of Energy (DOE), the U.S. Environmental Protection Agency, the U.S. Geological Survey, and the U.S. Department of Agriculture in the United States; the New Energy and Industrial Technology Development Organization in Japan; the United Nations Development Programme; and industrial companies such as Siemens and Caterpillar. Fan has helped various chemical, environmental, and energy companies overcome their technical challenges. He has published many refereed papers in different chemical and environmental engineering, energy, and chemistry journals. He is one of the highly cited researchers according to Web of Science. His recent NSF and DOE projects cover the areas of carbon capture, utilization, and storage; catalyzed solar-energy-driven biomass conversion; rare-Earth oxide extraction; reduction of rare-Earth oxides to rare-Earth metals; carbon fuel cells; and the production of chemicals, materials, and fuels from fossil resources.

MATTHEW FRY joined the Great Plains Institute in August 2021 as the state and regional policy manager, supporting the Carbon Management program. Fry has over 20 years of experience in natural resource management, regulation, and policy in both the public and private sectors. He served as a senior policy advisor to Wyoming Governor Matt Mead, where he focused on natural resource, energy, and carbon capture, utilization, and storage (CCUS) policy. Additionally, he developed and managed the Wyoming Pipeline Corridor Initiative, which is a project that authorized a statewide network of pipeline corridors in Wyoming that aimed to establish corridors on public lands dedicated for future use of pipelines associated with CCUS, enhanced oil recovery, and delivery of associated petroleum products. Fry earned a B.S. in biology and chemistry from Davis & Elkins College, and a master's degree in natural resource law from the University of Denver Sturm College of Law.

HAROUN MAHGEREFTEH is a professor of chemical engineering at the University College London and a fellow of the Institution of Chemical Engineers. His research spans all aspects of carbon capture, utilization, and storage (CCUS), particularly CO_2 pipeline safety and operational issues. In CCUS, he is the coordinator of several national and multinational collaborative projects, including the European Commission FP7 and H2020 projects, CO_2PipeHaz, CO_2QUEST, and C4U (total funding $32 million). Project highlights include the development of best practice guidelines for injection of CO_2 into highly depleted gas fields and the construction of the world's longest fully instrumented CO_2 pipeline rupture test facility located in Dalian, China. He is one of the two lead authors of the Zero Emission Platform report titled "A Trans-European CO_2 Transportation Infrastructure for CCUS: Opportunities & Challenges." The report is aimed at facilitating the development of a pipeline and ship infrastructure for transporting several million tonnes CO_2 per year captured from major regional industrial emitters for permanent offshore geological storage; considered as a key enabler for meeting net-zero-emissions target by 2030. His PipeTech computer program is routinely used by several major international corporations and legislative organizations for safety analysis of thousands of kilometers of pressurized pipelines across the globe.

EMANUELE MASSETTI is a technical assistance advisor at the Climate Unit of the Fiscal Affairs Department of the International Monetary Fund and an associate professor in the School of Public Policy of the Georgia Institute of Technology (on leave until December 2022), where he leads the Laboratory for Integrated Economics Engineering Environment Assessment and Policy. Previously he was a postdoctoral fellow at the Yale School of the Environment and a senior researcher at the Euro-Mediterranean Center on Climate Change. He contributed as the lead author to the Fifth Assessment Report of Working Group III of the International Panel on Climate Change. His research is on the economics of climate change, both mitigation and adaptation. He is the co-author of the Integrated Assessment Model WITCH and has authored or co-authored over 30 publications on climate change mitigation, adaptation, and impacts published in international peer-reviewed journals, including *Review of Environmental Economics & Policy*, *Energy Economics*, *The Energy Journal*, *Environmental and Resource Economics*, and *Climatic Change*. He holds a Ph.D. in economics from the Catholic University of Milan, Italy.

AH-HYUNG (ALISSA) PARK is the Lenfest Earth Institute Professor of Climate Change in the Departments of Earth and Environmental Engineering and Chemical Engineering at Columbia University. She is also the director of the Lenfest Center for Sustainable Energy. Her research focuses on sustainable energy and materials conversion pathways with emphasis on integrated carbon capture, utilization, and storage (CCUS) technologies addressing climate change. Park's group is also working on direct air capture of CO_2 and negative emission technologies including bioenergy with carbon capture and storage and sustainable construction materials with low carbon intensity. Park has received a number of professional awards and honors including the U.S. C3E Research Award (2018), the PSRI Lectureship Award at the American Institute of Chemical Engineers (AIChE) (2018), the American Chemical Society (ACS) Energy and Fuels Division Emerging Researcher Award (2018), the ACS Women Chemist Committee Rising Star Award (2017), and the National Science Foundation CAREER Award (2009). Park has also led a number of global and national discussions on CCUS, including the Mission Innovation Workshop on CCUS in 2017 and the National Petroleum Council CCUS Report in 2019. She is an elected fellow of the American Association for the Advancement of Science, AIChE, ACS, and the Royal Society of Chemistry.

JOSEPH B. POWELL is a fellow and former director of the American Institute of Chemical Engineers (AIChE) and served as Shell's first chief scientist–chemical engineering from 2006 until retiring at the end of 2020, culminating a 36-year industry career where he led research and development (R&D) programs in new chemical processes, biofuels, and enhanced oil recovery, and advised on R&D for the energy transition to a net-zero carbon economy. Powell is the co-inventor on over 125 patent applications (60 granted), has received AIChE, American Chemical Society, and *R&D Magazine* awards for innovation, service, and practice, and is the co-author of *Sustainable Development in the Process Industries: Cases and Impact* (2010). He chaired the U.S. Department of Energy's Hydrogen and Fuel Cell Technical Advisory Committee and was elected to the National Academy of Engineering (2021) after serving two terms on the National Academies of Sciences, Engineering, and Medicine's Board on Chemical Sciences and Technology. Other roles include guest editor of *Catalysis Today* on natural gas utilization, editorial board for *Annual Review of Chemical and Biological Engineering*, and cross-cutting technologies area lead and author for *Mission Innovation Carbon Capture Utilization and Storage* (2017). He currently advises in energy and chemicals and process development (ChemePD LLC). He received a Ph.D. from the University of Wisconsin–Madison (1984) and a B.S. from the University of Virginia (1978), both in chemical engineering.

ANDREA RAMÍREZ RAMÍREZ is a professor of low carbon systems and technologies at the Delft University of Technology, the Netherlands. Her research focuses on the evaluation of novel low-carbon technologies and the design of methodologies and tools to assess their potential contribution to sustainable industrial systems. Ramírez currently coordinates the research line on System Integration and Fair Governance of the Dutch project RELEASE, aiming to develop reversible large-scale energy storage based on electrochemical conversion of CO_2 into molecules. In 2018, she was awarded one of the largest scientific grants for individuals in the Netherlands to investigate the system impacts of using alternative raw materials such as CO_2, biomass, and waste in petrochemical industrial clusters. In the past 10 years, Ramírez co-coordinated the European project Environmental Due Diligence of Novel CO_2 Capture and Utilization Technologies, led the research line Techno-economic and Environmental Analysis of the Dutch R&D program Catalysis for Sustainable Chemicals from Biomass, and coordinated the program line Transport and Chain Integration of the Dutch R&D program for CO_2 Capture, Transport, and Storage. Ramírez holds a bachelor's degree in chemical engineering, a master's degree in human ecology, and a Ph.D. in industrial energy efficiency. She has authored or co-authored over 115 publications and is the editor-in-chief of the *International Journal of Greenhouse Gas Control*.

VOLKER SICK is the DTE Energy Professor of Advanced Energy Research and an Arthur F. Thurnau Professor at the University of Michigan, Ann Arbor. He leads the Global CO_2 Initiative at the University of Michigan that aims to reduce atmospheric CO_2 levels by transforming CO_2 into commercially successful products using technology assessment, technology development, and commercialization. His research focuses on accelerating deployments of CO_2-utilization technologies that will innovate existing infrastructure and manufacturing processes, thereby finding sustainable decarbonization solutions and continued access to required carbon-based products to help address the climate crisis. The author of numerous publications in both peer-reviewed and popular periodicals, his most recent awards and honors include the Royal Society of Chemistry Spiers Memorial Lecture Award (2021), the DTE Energy Professor of Advanced Energy Research (2019), and the President's Award for Distinguished Service in International Education (2018). He is a fellow of SAE International (2007) and a fellow of the Combustion Institute (2018). He received his doctorate in chemistry and habilitation in physical chemistry from the University of Heidelberg, Germany. He joined the University of Michigan as a professor of mechanical engineering in 1997.

B

Disclosure of Conflicts of Interest

The conflict of interest policy of the National Academies of Sciences, Engineering, and Medicine (http://www.nationalacademies.org/coi) prohibits the appointment of an individual to a committee authoring a Consensus Study Report if the individual has a conflict of interest that is relevant to the task to be performed. An exception to this prohibition is permitted if the National Academies determine that the conflict is unavoidable and the conflict is publicly disclosed. A determination of a conflict of interest for an individual is not an assessment of that individual's actual behavior or character or ability to act objectively despite the conflicting interest.

Makini Byron has a conflict of interest in relation to her service on the Committee on Assessing Carbon Utilization Infrastructure, Markets, Research and Development because of her employment at Linde, an industrial gas company that separates and purifies CO_2 and sells it to other companies for conversion into valuable products. The National Academies have concluded that the committee must include a member with current industry experience in managing the link between the commercial sources of the carbon dioxide—including the processes for capturing the carbon dioxide and the costs and quality of the carbon dioxide obtained—and the current and emerging markets for this carbon dioxide, including the quality requirements, cost requirements, and the potential quantities that might be utilized. As described in her biographical summary, Byron has extensive industry experience in understanding the innovation and costs of carbon dioxide capture and the use of this carbon dioxide in products. Byron has managed or participated in several U.S. Department of Energy (DOE)-funded projects for both the commercial engineering design and the scale-up demonstration of Linde's carbon capture technology developed with BASF. Her project-based knowledge also extends to the biological conversion of CO_2 to valuable products, the mineralization of CO_2 to cementitious material, and the application of supercritical CO_2 for lubrication and cooling. The National Academies have determined that the experience and expertise of Byron are needed for the committee to accomplish the task for which it has been established. The National Academies could not find another available individual with the equivalent expertise and breadth of experience who does not have a conflict of interest. Therefore, the National Academies have concluded that the conflict is unavoidable. The National Academies believe that Byron can serve effectively as a member of the committee, and the committee can produce an objective report, taking into account the composition of the committee, the work to be performed, and the procedures to be followed in completing the study.

Stephen Comello has a conflict of interest in relation to his service on the Committee on Assessing Carbon Utilization Infrastructure, Markets, Research and Development because of his technical consulting with Carbon Direct, a company that invests in carbon removal and utilization technologies, and his role as an external advisor to energy practice at the consulting firm Bain & Company. The National Academies have concluded that, given the study's focus on market opportunities for carbon dioxide–derived products and carbon utilization technologies, it is essential to have a committee member with current experience in financing methods, business models, and decision-making strategies that enable the development and deployment of clean energy technologies. As described in his biographical summary, Comello possesses a unique combination of technology and economic expertise. Comello integrates tools and approaches from engineering, finance, and systems analysis to develop methodologies for analyzing investments and innovations in low-carbon energy solutions. His expertise spans an array of business analytical skills, including environmental economics, decision analysis, life cycle assessment, and techno-economic evaluation for advanced clean energy technologies. As a technical adviser to Carbon Direct, Comello brings an understanding of the technological and organizational capabilities of startup companies in carbon utilization, which is critical for addressing the committee's task of determining how federal agencies can support small business to further the development and deployment of carbon dioxide–based products. The National Academies have determined that the experience and expertise of Comello are needed for the committee to accomplish the task for which it has been established. The National Academies could not find another available individual with the equivalent expertise and breadth of experience who does not have a conflict of interest. Therefore, the National Academies have concluded that the conflict is unavoidable. The National Academies believe that Comello can serve effectively as a member of the committee, and the committee can produce an objective report, taking into account the composition of the committee, the work to be performed, and the procedures to be followed in completing the study.

Ah-Hyung (Alissa) Park has a conflict of interest in relation to her service on the Committee on Assessing Carbon Utilization Infrastructure, Markets, Research and Development because of her equity in the startup company GreenOre CleanTech, LLC. Park is a co-founder of GreenOre, which focuses on carbon capture and process design, using carbon dioxide and other waste streams to generate valuable products. The National Academies have concluded that, given the rapidly accelerating developments in the science, engineering, and commercialization of carbon utilization technologies, it is essential to have a committee member with current experience in basic research activities and knowledge of the opportunities and processes for technology scale-up in this field. As described in her biographical summary, Park has an active research program spanning many topics relevant to the study, including CO_2 mineralization, materials for CO_2 capture and gas separations, chemical CO_2 conversion, and clean hydrogen production. In addition to her experience as an expert and leader in carbon capture and utilization research, her experience with GreenOre translating academic research into a startup company makes her expertise a critical addition to this committee. The National Academies have determined that the experience and expertise of Park are needed for the committee to accomplish the task for which it has been established. The National Academies could not find another available individual with the equivalent expertise and breadth of experience who does not have a conflict of interest. Therefore, the National Academies have concluded that the conflict is unavoidable. The National Academies believe that Park can serve effectively as a member of the committee, and the committee can produce an objective report, taking into account the composition of the committee, the work to be performed, and the procedures to be followed in completing the study.

Joseph Powell has a conflict of interest in relation to his service on the Committee on Assessing Carbon Utilization Infrastructure, Markets, Research and Development because of his stock in Royal Dutch Shell, plc. The National Academies have concluded that the committee must include a member with recent experience and expertise in the chemical fuels industry with an understanding of the industrial and process engineering involved in producing such fuels and potentially adapting existing infrastructure for utilizing captured carbon dioxide in products. As described in his biographical summary, Powell has had extensive experience in development, scale-up, and commercialization of existing and new technologies. He also has industrial systems expertise that is vital to address the committee's task of assessing infrastructure and research and development needs to support a future circular carbon economy. The National Academies have determined that the experience and expertise of Powell are needed for the committee to accomplish the task for which it has been established. The National Academies could not find another available individual with the equivalent expertise and breadth of experience who does not have a conflict of interest. Therefore, the National Academies have concluded that the conflict is unavoidable. The National Academies believe that Powell can serve effectively as a member of the committee, and the committee can produce an objective report, taking into account the composition of the committee, the work to be performed, and the procedures to be followed in completing the study.

C

Information-Gathering Activities

January 11: Open Session with the U.S. Department of Energy (DOE) and Congressional Sponsors

DOE's Office of Fossil Energy and Carbon Management
- Emily Grubert, Deputy Assistant Secretary for Carbon Management
- Darin Damiani, Senior Program Manager
- Amishi Kumar, Carbon Utilization Program Manager

DOE's Office of Energy Efficiency and Renewable Energy
- Christy Sterner, Technology Manager for Advanced Algal Systems Program

DOE's Office of Science
- Bruce Garrett, Director of Chemical Sciences, Geosciences, and Biosciences Division
- Todd Anderson, Director of Biological Systems Science Division

Congressional Staff
- Adam Rosenberg, Subcommittee Staff Director, U.S. House Science, Space, and Technology Committee
- Luke Bassett, Senior Professional Staff Member, U.S. Senate Energy and Natural Resources Committee
- Armando Avila, Senior Professional Staff Member, U.S. Senate Energy and Natural Resources Committee

March 1–3: Carbon Utilization Information-Gathering Webinar Series

Day 1, March 1—CO_2-Derived Products
- Marcius Extavour, XPRIZE, moderator
- Rahul Shendure, CarbonBuilt
- Karl Haider, Covestro
- Julio Friedman, Columbia University, moderator
- Jennifer Holmgren, LanzaTech
- André Bardow, ETH Zürich
- Cathy Tway, Johnson Matthey

- Jean-Paul Lange, Shell
- Bill Tumas, National Renewable Energy Laboratory, moderator
- Etosha Cave, Twelve
- Ed Rightor, American Council for an Energy-Efficient Economy
- Gay Wyn Quance, Solid Carbon Products

Day 2, March 2—CO_2 Utilization Infrastructure
- Michael Drescher, Equinor
- Jesse Jenkins, Princeton University
- Marcius Extavour, XPRIZE
- Geoff Holmes, Carbon Engineering
- Jennifer Holmgren, LanzaTech
- Robert Niven, Carboncure
- Kristjana Kristjánsdóttir, Carbon Recycling
- Brett Perlman, Center for Houston's Future
- Bilal Ahmad, Northern Endurance Partnership
- Nico de Meester, Porthos Project

Day 3, March 3—Policy, Regulatory, and Societal Considerations for CO_2 Utilization
- Al Collins, Oxy
- Kevin Poloncarz, Covington & Burling, LLP
- Keith Tracy, Elysian
- Linda Daugherty, Department of Transportation
- Sheila Olmstead, University of Texas, Austin
- Vernice Miller-Travis, Miller-Travis & Associates
- Deepika Nagabhushan, Carbon Direct
- Patricia Romero-Lankao, National Renewable Energy Laboratory

April 29: Open Session with DOE's Office of Clean Energy Demonstrations

DOE's Office of Clean Energy Demonstrations
- Kelly Cummins, Acting Director and Principal Deputy Director
- Todd Shrader, Deputy Director for Project Management

D

Acronyms and Abbreviations

ACER	European Union Agency for the Cooperation of Energy Regulators
ACTL	Alberta Carbon Trunk Line
ANL	Argonne National Laboratory
ARI	Advanced Resources International
ATR	autothermal reforming
AWARE-US	Available Water Remaining for the United States
BBL	billion barrels of oil
Bcf/d	billion cubic feet per day
BECCS	bioenergy with carbon capture and storage
C1	one-carbon chemicals and materials
C2	two-carbon chemicals and materials
C2+	multi-carbon chemicals and materials
C2ES	Center for Climate and Energy Solutions
CAA	Clean Air Act
CATF	Clean Air Task Force
CBA	cost-benefit analysis
CC	carbon capture
CCS	carbon dioxide capture and storage
CCU	carbon dioxide capture and utilization
CCUS	carbon dioxide capture, utilization, and storage
CDR	carbon dioxide removal
CEQ	Council on Environmental Quality
CO_2e	carbon dioxide equivalents
CREZ	Competitive Renewable Energy Zone
CRI	Carbon Recycling International
CSIRO	Commonwealth Scientific and Industrial Research Organization
CU	carbon dioxide utilization

CUP	conditional use permit
CWA	Clean Water Act
DAC	direct air capture
DDP	Deep Decarbonization Pathways
DOC	direct ocean capture
DOE	U.S. Department of Energy
DOE-BETO	U.S. Department of Energy Bioenergy Technologies Office
DOE-EERE	U.S. Department of Energy Office of Energy Efficiency and Renewable Energy
DOE-FECM	U.S. Department of Energy Office of Fossil Energy and Carbon Management
DOE-HFTO	U.S. Department of Energy Hydrogen and Fuel Cell Technologies Office
DOE-NE	U.S. Department of Energy Office of Nuclear Energy
DOE-OCED	U.S. Department of Energy Office of Clean Energy Demonstrations
DOE-OEID	U.S. Department of Energy Office of Economic Impact and Diversity
DOS	U.S. Department of State
EASAC	European Academies Science Advisory Council
EC	European Commission
EEA	European Environment Agency
EFI	Energy Futures Initiative
EGR	enhanced gas recovery
EIA	U.S. Energy Information Administration
EIGA	European Industrial Gases Association
EIR	environmental impact report
EIS	environmental impact statement
EOP	Executive Office of the President
EOR	enhanced oil recovery
EPA	U.S. Environmental Protection Agency
ESA	Endangered Species Act
EU	European Union
FCHEA	Fuel Cell and Hydrogen Energy Association
FEED	front-end engineering design
FERC	Federal Energy Regulatory Commission
FTS	Fischer-Tropsch synthesis
GHG	greenhouse gas
Gt	gigatonne
GW	gigawatt
GWP	global warming potential
H_2S	hydrogen sulfide
HGL	hydrocarbon gas liquid
IDEALHy	Integrated Design for Efficient Advanced Liquefaction of Hydrogen
IEA	International Energy Agency
IEAGHG	IEA Greenhouse Gas R&D Programme
IIJA	Infrastructure Investment and Jobs Act
IPCC	Intergovernmental Panel on Climate Change
IRA	Inflation Reduction Act

APPENDIX D

IRENA	International Renewable Energy Agency
ITIF	Information Technology & Innovation Foundation
IWG	International Working Group
kWh	kilowatt hour
LADWP	Los Angeles Department of Water and Power
LBNL	Lawrence Berkeley National Laboratory
LCA	life cycle assessment
LEP	Labor Energy Partnership
LNG	liquefied natural gas
LPG	liquefied petroleum gas
MCC	Mercator Research Institute on Global Commons and Climate Change
MEA	monoethanolamine
MMT	million metric tons, million tonnes, megatonnes
MP	methane pyrolysis
Mtpa	million tonnes per annum
MWhe	megawatt-hours of electricity
NEPA	National Environmental Policy Act
NETL	National Energy Technology Laboratory
NGA	Natural Gas Act
NJDEP	New Jersey Department of Environmental Protection
NO_x	nitrogen oxides
NPC	National Petroleum Council
NPDES	National Pollutant Discharge Elimination System
NPMS	National Pipeline Mapping System
NSR	New Source Review
PARC	Palo Alto Research Center
PEM	proton exchange membrane
PHMSA	Pipeline and Hazardous Materials Safety Administration
POx	partial oxidation
PPA	power purchase agreement
PSD	Prevention of Significant Deterioration
PTC	production tax credit
PV	photovoltaic
R&D	research and development
RD&D	research, development, and demonstration
REPEAT	Rapid Energy Policy Evaluation and Analysis Toolkit
SMR	steam methane reforming
SOx	sulfur oxides
TEA	techno-economic assessment
TRL	technology readiness level
TWh	terawatt hour

WE	water electrolysis
WTI	West Texas Intermediate crude oil
UNFCCC	United Nations Framework Convention on Climate Change
USACE	U.S. Army Corps of Engineers
USCA	United States Climate Alliance
USD	U.S. dollar
USFWS	U.S. Fish & Wildlife Service
ZEP	Zero Emissions Platform